本土设计 Ⅲ

LAND-BASED RATIONALISM Ⅲ

崔愷 著
CUIKAI

中国建筑工业出版社
China Architecture & Building Press

序

崔院士邀请我为《本土设计Ⅲ》写序言，既感荣幸也有些许惶恐。我和崔总同年出生，是相熟多年的朋友。他作为"文革"后成长起来的中国建筑界最具代表性的业界翘楚，国内外享有盛誉，很多作品早已蜚声远近。他的不少作品也是我学习参访的对象，我曾经在1990年代中叶研究工业建筑遗产保护改造，其时专门参观了他设计的北京外研社项目，因为其中包括了对原印刷厂房的改造利用；前几年又在江苏昆山崔总设计的砖窑厂改造项目中参加了江苏乡村振兴的会议，苏州火车站等也是我非常熟悉的建筑。这几年因第11届江苏省园艺博览会和第13届中国（徐州）国际园林博览会项目，我们又有了在同一场地环境合作设计的机会。此次的《本土设计Ⅲ》是对崔总及团队近十年作品的系统总结和整体梳理，很高兴有机会第一时间看到此书的预印稿，先睹为快。

本书印象

如果说《本土设计》（2009）在建筑理论和方法上带有某种试错证伪意义的开拓性，那么《本土设计Ⅱ》（2013）则进一步探索、深化和延展了"本土设计"的核心理念并完成了理论构型，《本土设计Ⅲ》（2023）则又在方法、判断力、沟通力、风格等方面完成了本土设计理论方法与工程实践"互鉴"的系统建构，设计领域则更加突出了城市设计、城市更新、"美丽中国"等方面的探索，从一个侧面可以体会到十年来国家城镇化背景下城市发展模式的改变，以及由此引发的人们对人居环境高质量发展的认识演进。从2013年中央城镇化工作会议提出"望得见山、看得见水、记得住乡愁"，到2015年中央城市工作会议强调城市设计的重要性，再到每年"中央1号"文件对"三农"问题的关注，新近强调低碳绿色、生态保护和城乡特色保护，崔总总能从时代发展趋势中看到建筑师的责任和职业发展中的机遇，并以自己积极能动的实践方式去回应国家和社会城乡建设中的重大需求。

"本土设计"的英文译名是 Land-based Rationalism，从中文翻译字面上看，就是基于本土的理性主义，实际上就是一种趋近成熟而凝练出来的设计理念和价值系统。这种设计理念和价值系统贯穿了《本土设计Ⅲ》全书文字和实践作品内容的分类：设计方法、设计判断力、沟通力、风格、象征、城市设计、城市特色风貌营造及城市更新。

个人观察

在国家级大型设计院工作，与主要侧重前期原创的设计机构和事务所有所不同，国际上也是如此。大体来说，中国院（中国建筑设计研究院）为代表的国内大型设计院与SOM、KPF、GMP、HKS等国外顶尖设计机构突出强调的是机构整体的协同设计力量、攻坚克难的技术优势和大型复杂建筑错综复杂设计问题的解决。在《本土设计Ⅲ》中，少数作品是我比较熟悉的，有些还在我们合作设计的场地内。去年我去西藏拉萨，在住的酒店不远看到一幢设计精良的高层建筑，当时我就对同事说这个建筑设计比例、尺度、色彩和藏式元素表达很有功力，这次看到书稿才知道原来是崔院士团队做的京藏交流中心。不过，更多的作品我还没有机会现场品鉴过，通过这本书也学到了很多。

《本土设计Ⅲ》记载的作品设计团队组成人员众多，崔院士根据不同的项目类别和大设计院的机构组织特点，带领了一支规模达数十人的团队开展工作，这让我想起，"传、帮、带"是中国大型建筑设计院一贯的优良传统。以中国院为例，不久前就办过一个建设部设计院（中国院前身）的发展历史展览，我还听过崔院士专门的讲解性报告；中国院还组织李兴钢、关飞等多位青年才俊解读老一辈大家的作品并交流认识体会，并出版了《重读经典——向前辈建筑师致敬》一书。这种传承的力量给我留下深刻影响。

建筑创作

崔总是一位"建筑感"特别好的建筑师，"建筑感"是我对崔总作品的第一层解读。

"建筑感"保证了他和团队的作品具有持久稳定的精良水准和审美品质，其中相当一部分作品曾经荣获各类奖项。崔总曾受到严谨求实的建筑学学习训练，是当年天大建筑系的"学霸"，获得过全国第一届大学生设计竞赛一等奖。他当年放弃留校任教，立志设计创作，并在深圳、北京先后设计了当年的建筑佳作，包括我前面提到的北京外研社建筑群等。这种"建筑感"具体包括了崔总所述的设计方法、设计判断力、风格象征的拿捏、城市设计概念等。

近年在多类型的建筑设计和城市设计实践中，我有一个切身的感受体会，就是寻找到既能表达设计理念、又能与相关领导、设计甲方和社会大众形成共识和共情的那一个设计终结的"纳什均衡"点最为重要，也即是崔院士已经具备并实践多年的那种高级的"建筑感"。同样重要的是，包括"本土设计"在内的经典建筑学本体价值内涵的工程性实现，恰恰是最能抵御当下声名鹊起的以GPT为代表的人工智能集群且其尚无法企及的设计内容。以往在设计方案阶段，决定方案的因素大多来自美学冲击力和形态功能合体的逻辑判断方面，而这方面在计算性设计兴起后可能会有所衰微。但在真实场地和场景中，具身在场的设计方法诠释，如书中写到的向领导和业主讲解江苏建筑职业技术学院图书馆为什么斜撑比较合理的案例，及判断力和沟通交流，是人工智能尚无法达到的。事实上，建筑具有多维度交织和错综复杂的价值构成，需要设计者推荐或者最终决策，选取"纳什均衡"相对最优解。因为只能有一个方案能得以实施，如果自己钟爱的不止一个选择，也只能是"零和博弈"，择一而终。

熟谙建筑设计"基线"，勇于超越过往经验，不断实现"创作边界"的突破，是我对崔总作品的第二层解读。

通常，建筑师工作具有"基线"和"边界"。坚守建筑本体的"基线"，也就是坚守专业基础。崔总接受过系统性的建筑专业教育，通晓建筑设计逻辑，能够很好把握如何按照行业通行的规范和规则设计建筑，并能将其在保证质量、业主认可的前提下将其完美建成。书中写到本土设计具有普遍意义，是一种基本理性的方法论，是一种近乎"公理"的方法论，在我看来，或许就可看成是某种"经典性"，如天津大学新校区主楼设计的经典性布局和形象特征表征的就是这种"经典性"。然而，我们又能清晰看到崔总作品对设计"边界"的持续探索和跨越。"边界"概念一般理解比较宽泛、也很模糊，绝大多数建筑师都是在边界内做事情，但是，创新和跨越往往也需要在特定的设计场合去探索"边界"的"临界"和"超界"的多种可能性。

我一直认为"边界"的探索不仅是有理想的建筑师实现"应然理想"目标的必由之路，而且可以激发创意，探索未知。良好的跨界知识思维和综合创造的能力对于建筑师很重要，在崔总系列作品中，我们可以看到很多这样属于新的"边界"的探索和知识进阶的尝试，如新近的天府农业博览园主展馆、北京世界园艺博览会中国馆等在绿色建筑和节能方面的突破；山东荣成市少年宫对地景建筑和清水混凝土使用的尝试；运用集装箱单元完成的雄安中心项目的探索等。我曾经接待和参访过一些国际级设计大师，深感他们关注的领域和拥有的专业素养早已超出建筑学的领域。不过，对于大多数人来说，跨界也是有极限和天花板的，不是为跨而跨，而应该是解题导向和进阶性的合理跨界。建筑师群体超越"边界"和"跨界"首先应该建立在"基线"合格的基础上（个人认为，GPT产生的设计现在还不能应对工程落地性的实战，应该还没有具备"基线"要求），崔总书中曾经谈到关于建筑运用象征符号的三个要点，也谈到了不喜欢内外脱节的"表皮设计"，如果只看"外表"，大概率会被人工智能所取代。我个人十分欣赏先锋和边缘的设计探索，但不喜欢过度的包装策划、但随即成为过眼烟云般"一夜成名"的所谓"明星"。建筑还是一个真实建立在物理空间、地形场地、使用功能、依法合规和集体审美基础上的"社会系统的人工物设计"。

《建筑学报》约我写学习杨廷宝先生体会文章时，我曾写道，"作为建筑巨匠，他的设计总是能够游走在多极设计目标诉求和工程实现之间，在文化、环境、技术、造价中取得合理平衡，通过诸多小的不确定性博弈找到最后的确定性——房屋建成"。我觉得，不少设计机构的优秀建筑师都具有这样的品质，如崔总经常提到的戴念慈先生。崔总的实践成果也完全可以这样评价。

在地性思想理念、设计体系化和"本土"价值偏好，是我对崔总作品的第三层解读。

"一方水土养一方人"，崔总从不掩饰他对中国本土文化的眷恋，崔总的设计因地制宜，不少设计创意都由场地调研感悟所激发。他在贵州兴义市做镇村规划设计时曾提出过"蔓藤城市"的构思，并落到城市和建筑设计的方案中；但同时他具有开阔的国际视野，在设计访谈中多次谈到库哈斯等建筑大师的作品对他设计的启发。"本土设计"不是指设计的出发点只有"本土",而是他大多数建筑设计的真实场景具有中国"在地性"(locality)。这种"在地性"既可以是一种本土眷恋的乡愁(vernacular)和集体记忆(collective memory),也可以是特定设计场地自然地形、地貌和风土之于设计创作的启示。它的作品或厚重敦实(敦煌市博物馆)、或轻盈灵动(南浔城市规划展览馆、徐州园博园儿童友好中心等),并不拘泥于某种稳定和确定的个人标签风格特征。我自己也十分看重设计特定在地性的某种呈现和表达。

"本土"价值不仅在于特定场地自然条件和文化风土的解读,那只是建筑师单向解读和认识的对象,还有一部分重要内容就是"人",其中当然首先是要考虑社会大众对设计工程建成的期待和需求,但同时业主领导、委托方及项目出资人同样也具有对设计进行点评和批评建议的权利,所以,崔总特别强调"沟通力"的重要性。建筑工程实践是一个典型的多目标交集而求其相对最优的达成过程。建筑设计过程不是简单的"专家"面对"外行"的情形,应尽可能地增加过程组织的"透明度"。设计伊始,建筑师和业主方及决策领导都应该彼此充分沟通交流,双方都需要在设计过程中学习新的知识,甚至包括提出设计任务书调整和完善的建议,崔院士常用的就是"本土设计"的一系列理念和设计思路,然后通过一个多目标优化互动协调的过程,最终进行基于成功概率的主观决策判断,实现设计创意。

建筑师责任

当下,中国正处在倡导生态文明建设、中华优秀文化传承以及数字科技快速发展的现代化发展阶段。建筑师群体正在越来越广泛地涉及绿色建筑、城市更新、乡村振兴、城市设计等领域的知识展拓及实践,崔院士团队近年已经对此进行了卓有成效的探索。如他书中提到的绿色建筑需要做"减法"——"减少建用地、较少拆房、减少侵占自然、减体量、减用能空间和时间、减少玻璃幕墙、减结构、减少装饰、减运行成本"等,相关成果还编成了建筑师绿色设计的导则手册。我认为,这种基于建筑师设计思维的绿色低碳理解对我国今后建筑设计价值导向具有很大的启发。建筑师不仅需要追求视觉之美、环境之序、场所之真、文化之谐,还要追求伦理之善等隐含的世界可持续发展的价值要求。书中介绍了崔总本土设计七个方面的目标和策略:文脉传承、遗产保护、城市更新、风景融入、绿色导向、生活引领和乡村振兴,清晰呈现了建筑师责任的实践探求。

最近几年,崔院士担任了中国注册建筑师管理委员会主任,我也担任了该委员会的考试组组长。每年都有几次与崔院士和全国众多设计院老总或院领导在命题、改卷及大纲优化等方面的讨论见面的交流机会。近年,他又在北京建筑大学开始系统教授"本土设计"的研究生课程,由此高校学子与大师之间又有了直接面对面的学习交流机会。他跟我说,在学生对他授课的反馈中也了解了更多关于年轻一代建筑人才培养的意义和重要价值。

崔院士在我的印象中特别忙,即使我们中国工程院在北京会议中心开会的时候,他也经常利用中午和晚上的时间处理各种工程设计问题,这种敬业精神值得我学习。

总体而言,《本土设计Ⅲ》整本书理论体系脉络清晰、设计方法因地制宜、建筑实践形神兼备,每个设计作品均有几个"关键词"引导阅读。该书基于崔院士团队多年来开展的大量且多类型的建筑设计工程实践优秀案例,提出了一系列兼具设计理念、场地解读、难题应对、技术选择和内涵诠释等建筑界业内普遍关注的设计策略、技术方法和工程实施要点,对于建筑师群体,也包括规划师以及景观设计师等群体是一本非常有价值的专业宝典和参考书;同时,对于关注正在成为热点的城市更新和城乡人居环境高质量发展的大众也是一本图文并茂、虚实兼备、可以帮助专业认识进阶的科普论著。

王建国
2023年4月

PREFACE

When Academician CUI Kai invited me to write the preface of this book, I felt both honored and somewhat unsure if I was fully qualified. We are of the same age and have been friends for many years. As one of the most prominent architects emerging after the Cultural Revolution, CUI Kai is renowned both domestically and abroad, with many well-known projects, a large portion of which were great reference of my study. In the mid-1990s, while I was carrying out research on industrial heritage preservation and renovation, I visited the Office Building of Foreign Language Teaching & Research Press, featuring renovation and reutilization of a printing house. A few years ago, I attended a conference on rural revitalization in a brick factory renovation in Kunshan, Jiangsu, which was renovated by CUI's team. I'm also quite familiar with a series of his other projects such as Suzhou Railway Station. Thanks to the Jiangsu Garden Expo and the Xuzhou Garden Expo, I had the opportunity to work in the same site with CUI Kai during the past few years. I view this book as a systematic summary and overall review of his works over the past decade, and I am delighted to read the pre-print version of the book before its release.

My Impression of the Book

If *Native Design* (2009) was of pioneering significance for its trial-and-error approach in terms of architectural theory and methods, *Land-based Rationalism II* (2013) explored, deepened, and extended the core concept of "land-based rationalism design" and completed its theoretical framework. *Land-based Rationalism III* (2023) has completed the systematic framework where theory, method and engineering practice are integrated. The book has revealed his new efforts in urban design, urban renewal, and the beautiful China campaign, through which we can see the changes of the urban development model during the urbanization process over the past decade, as well as people's new understanding of high-quality development in livable environments. From the Central Urbanization Work Conference's proposal of "Make real mountains and waters seen by people with nostalgia at heart" in 2013, the emphasis on urban design at the 2015 Central Urban Work Conference, and the focus on issues relating to agriculture, rural areas, and rural people that are mentioned annually in the Central Government's No.1 Document, to the government's recent emphasis on green development, ecological protection, and the protection of urban and rural characteristics, Academician CUI always sees through those changes the responsibility and opportunities of architects, and responds to the major needs of urban and rural construction through his proactive practices.

His design approach, translated as "Land-based Rationalism", literally means rationalism based on the land. Actually, it is both a design concept and a value system that were generated through its process of maturity. These are all reflected in the contents of *Land-based Rationalism III*, classified as design methods, design judgment, communication, style, symbolism, urban design, urban characteristics creation and urban renewal.

Personal Observation

In contrast to design firms or studios that mainly focus on early-stage conceptual design, national-level large design institutes such as CADG (China Architecture Design & Research Group), as well as internationally renowned firms such as SOM, KPF, GMP and HKS, emphasize the coordinative design capability of the design team, the technical advantages in overcoming difficulties, and the solution to complex problems. Among the works in *Land-based Rationalism III*, some are familiar to me, and some are even on the same site of my projects. Last year, I saw a well-designed high-rise building not far from the hotel where I was staying at Lhasa, and I told my colleagues that the proportions, scales, colors, and expression of Tibetan elements of the building were very skillfully designed. It turned out that it was the Beijing-Tibet Communication Center designed by CUI's team. Though I haven't been to a large number of his works, I have learned a lot through this book.

In *Land-based Rationalism III*, names of many team members have been recorded, meaning that Academician CUI leads a team of dozens of people. This reminds me that mentoring is a long-standing tradition in large architectural design institutes in China. For example, CADG once held an exhibition on its history of development. I have also heard Academician CUI's lecture in explanation of the exhibition. Recently, young talents of CADG such as LI Xinggang and GUAN Fei studied architectural works of earlier generations of architects and shared their understanding. This sort of inheritance has left a deep impression on me.

Architectural Creation

CUI is an architect with a very good "sense of architecture", which is my first impression of him.

This "sense of architecture" has ensured the stable quality of Academician CUI's works, of which a considerable part are award-winning ones. CUI has received high-quality architectural education and was a top student in the architecture department of Tianjin University. He won the first prize in the National College

Student Design Competition, and gave up teaching to devote himself to architectural creation. He has designed a series of excellent works in Shenzhen and Beijing, including buildings in Beijing Foreign Language Teaching & Research Press mentioned above. This kind of "sense of architecture" refers to design methods, design judgment, style and symbolism, as well as urban design concepts.

Over my architectural and urban design practices during the past years, I have a personal feeling that what is most important lies in finding the "Nash equilibrium" point of the final design that can express the design concept and form a consensus and empathy with authorities, designers and the public. This is a sort of advanced "sense of architecture" that Academician CUI has possessed and practiced for many years. Equally important is the engineering implementation of the essential value connotation of classical architecture, including land-based rationalism, which is resistant to AI clusters represented by GPT, as AI cannot accomplish such design content. In the past, factors that determined the design scheme mostly came from aesthetic impact and logical judgment of the form-function combination, which might have declined after the rise of computational design. However, in real sites and scenes, the interpretation of the design method embodied in the scene, such as explaining why Library of Jiangsu Jianzhu Institute, have diagonal struts, as is described in the book, as well as judgment and communication, are essential. In fact, architecture is the composition of multi-dimensional interweaving and complex values, and architects need to recommend or decide the relative optimal solution to "Nash equilibrium" for the final decision, for there can only be one choice for implementation, otherwise the design will be a zero-sum game.

Being familiar with the "baseline" of architectural design and daring to go beyond past experiences to make breakthroughs of the "creation boundary", this is my second interpretation of Academician CUI's work.

Usually, architects' work has a "baseline" and a "boundary", and adhering to the "baseline" is fundamental to professional practices. Academician CUI, who has received systematic architectural education, understands architectural design logic very well, allowing him to grasp how to design buildings according to industrial norms and rules, and ensure the building's quality while gaining clients' recognition. The book mentions that land-based rationalism has universal significance as a sort of basic rational methodology. To me, it can be seen as a kind of "classical" manifested by the layout and characteristics of Main Building of Tianjin University New Campus. We can also see CUI's continuous exploration and crossing of the design boundary. The concept of boundary is generally broad and vague, with the vast majority of architects working within the boundary. However, innovation often requires exploring marginal possibilities, or even transcending the boundary in specific design situations.

I have always believed that the exploration of the "boundary" is not only the necessary way for architects to achieve the goal of "ideal", but also can inspire creativity and exploration of the unknown. Interdisciplinary knowledge and comprehensive creativity are essential for architects. In CUI's series of works, we see many attempts to explore new "boundaries" and advance knowledge. Examples include breakthroughs in green design and energy conservation in recent projects such as Main Exhibition Hall of Tianfu Agricultural Expo Park and China Pavilion at Beijing International Horticultural Expo 2019, the experimentation with landscape architecture and fair-faced concrete at Rongcheng Youth Activity Center in Shandong, as well as the exploration of the Xiong'an Civic Service Center Enterprise Office Area's container units. I have received or visited some international design masters and felt that their concern and expertise have already gone beyond the field of architecture. However, for most people, crossing boundaries also has its limits. It should not be about crossing just for the sheer sake of itself; it should be guided by problem-solving and advanced thinking. Architects' transcendence and crossing of boundaries should first be based on a qualified "baseline" (in my personal opinion, GPT-generated design cannot yet meet the practicality of engineering, so it does not meet the "baseline" requirement). CUI mentioned three key points about the symbolic representation in architecture in his book, and also talked about his disapproval of envelope design with detached exterior and interior spaces. If we only focus on the "appearance", we are very likely to be replaced by AI. Personally, I appreciate pioneering explorations, but dislike the so-called "meteors" who become famous overnight with excessive plotting, but soon disappear. Architecture is still an "artificial object design of social system" based on physical space, terrain, functional use, legal compliance, and collective aesthetic beliefs.

When the Architecture Journal invited me to write an article about my experience learning from Mr. YANG Tingbao, I wrote, "As a master architect, he always balanced between multiple design objectives and engineering implementation, achieving reasonable balance in culture, environment, technology, and cost. Through tackling with many small uncertainties, he finds the final certainty - the completion of the building." I think many excellent architects possess these qualities, such as DAI Nianci mentioned frequently by CUI. And CUI's architectural practices can also be described with those above-mentioned words.

The third level of my interpretation of Academician CUI's works is the systematization of his theory and practice of locality, as well as his preference to "land-based" values.

As the old saying goes, "Each place has its own way of supporting its own inhabitants." Academician CUI never disguises his affection for Chinese culture.

His designs are adaptive to the local environments, and many design ideas are inspired by his observation of the site. For the planning and design of the Xingyi in Guizhou Province, he proposed the concept of "ivy city", which was incorporated into both the urban and architectural designs. At the same time, he also possesses a broad international perspective and has mentioned many times in design interviews the works of masters such as Koolhaas that have inspired his designs. "Land-based rationalism" does not only mean that the starting point for design is based on the land, but that most of his architectural designs have a Chinese "locality", which can be a kind of vernacular and collective memory of local attachment, or the inspiration from the terrain, landscape, and local culture of a specific design site. His works may be solid and substantial (Dunhuang Museum) or light and dexterous (Nanxun Planning Exhibition Hall, Children-friendly Center of Xuzhou Garden Expo, etc.), of which none is limited to a certain personal style. I myself also highly value the expression of the specific localization.

"Land-based" values not only involve the interpretation of natural features and cultural customs of a specific site, which is a sort of one-way interpretation. However, there is also a crucial part that concerns "people", with the expectations and needs of society as the top priority. At the same time, clients and investors also have the right to give feedback and criticism. Therefore, CUI emphasizes the importance of "communication". Architectural practice is a typical process where multiple goals coexist and a relatively optimized result is pursued. The process is not a simple situation of "experts" facing "laymen". Instead, more transparency should be added to the process. From the beginning of design, architects and clients should communicate effectively, as both sides need to gain new knowledge during the design process, or even adjust the design brief. CUI often starts with a series of concepts of "land-based rationalism", and then through a coordinative process of multi-objective optimization, he finally makes a subjective decision based on probability of success to achieve the design concept.

Architects' Responsibility

At present, China is in an era of promoting ecological civilization construction, outstanding Chinese culture, and digital technology. Architects are expanding their knowledge in fields such as green building, urban renewal, rural revitalization, and urban design. In recent years, Academician CUI's team has made successful explorations in these fields. As mentioned in his book, green building requires "subtraction" - "to reduce land for construction, demolish fewer buildings, curb the occupation of nature, reduce the volume of buildings, energy-consuming spaces and practices, limit the usage of glazed walls, structures and decoration, and control operating costs", with relevant outcomes compiled into a manual serving as architects' green design guidelines. I think this kind of green and low-carbon understanding from architects' perspective has great significance for the value orientation of future architectural design in China. The pursuit of architects should not only focus on a building's beauty, environmental sequence, authenticity of place, and cultural harmony, but also the implicit values of sustainable development such as ethical goodwill. In this book, seven goals and strategies for land-based rationalism are introduced: cultural inheritance, heritage protection, urban renewal, landscape integration, green orientation, lifestyle guidance, and rural revitalization system, through which the exploration for architects' responsibility are clearly presented.

In recent years, Academician CUI has been serving as the director of the China Registered Architects Association, and I have been the head of the examination group of the association. So every year, I have several opportunities to meet with Academician CUI and design institute leaders and discuss topics related to the association. In recent years, Academician CUI has begun to teach the graduate course of "Land-based Rationalism" at Beijing University of Civil Engineering and Architecture, providing students with opportunities of face-to-face exchanges with design masters. He told me that he has also learned more about the importance of cultivating young generation of architects through students' feedback on his lectures.

In my impression, Academician CUI is always very busy. Even when we were gathering at the China National Convention Center in Beijing, he would often use the lunch break and evening time to deal with various design issues. His professionalism is worth learning from.

Overall, *Land-based Rationalism III* has shown a clear theoretical system, adaptive design methods, and building practices with united form and spirit. Each project has several "keywords" to guide the readers' understanding. Based on a large number and diverse types of excellent architectural practices by Academician CUI's team over the years, this book proposes a series of design strategies, technical methods, and implementation points that cover design concepts, site interpretation, problem solving, technology selection and connotation interpretation. It is a valuable guidebook and reference for architects, planners and landscape designers. It is also an illustrated educational book that can serve a broad range of readers interested in urban renewal and high-quality development of the urban and rural living environment.

WANG Jianguo
April, 2023

目　录

序 *002*

引子——本土设计方法论初探 *012*

设计方法
关于方法 *016*
本土设计方法思维导图 *017*
关于场地踏勘 *020*
关于判断力 *021*
关于沟通 *023*

文脉传承
南浔城市规划展览馆 *026*
东北大学浑南校区图书馆 *038*
昆山西部医疗中心一期 *048*
敦煌市公共文化综合服务中心 *060*
京藏交流中心 *068*
铁道游击队纪念馆 *076*
大同市博物馆 *086*
关于象征 *095*

遗产保护
江苏园博园主展馆及傲图格精选酒店 *098*
曲阜鲁能JW万豪酒店 *120*

城市更新
武汉大学城市设计学院教学楼 *130*
重庆市规划展览馆 *142*
昆山大戏院 *152*
宝鸡文化艺术中心 *160*
中国大百科全书出版社办公楼改造 *170*
关于城市之一：城市设计的维度与视角 *176*

关于城市之二：城市风貌的特色营造 *179*

风景融入
日照市科技馆 *182*
日照科技文化中心 *196*
兰州市城市规划展览馆 *206*
崇礼中心 *214*

生活引领
天津大学新校区主楼 *228*
北京邮电大学沙河校区 *240*
清华大学深圳国际研究生院一期 *248*
深圳万科云城 *256*
济南舜通大厦 *262*
青岛上合之珠国际博览中心 *268*
关于风格 *277*

乡村振兴
昆山西浜村昆曲学社及乡村工作站 *280*
昆山锦溪祝家甸砖厂改造 *290*
昆山小桃源 *298*

评论·访谈
厚土重本　大地文章——崔愷和他的"本土设计" *310*
面向中国本土的理性主义设计方法——崔愷院士访谈 *312*
大地生长——崔愷的敦煌"本土设计"建筑实践 *318*
方向与方法——源于本土设计实践的谈话 *324*

后记 *348*
附注 *350*
年表 *351*

CONTENTS

PREFACE *005*

INTRODUCTION: PRIMARY EXPLORATION OF LAND-BASED RATIONALISM METHODOLOGY *013*

DESIGN METHOD
ABOUT METHOD *016*
MIND MAPPING OF LAND−BASED RATIONALISM DESIGN APPROACH... *017*
ABOUT SITE SURVEY *020*
ABOUT JUDGEMENT *021*
ABOUT COMMUNICATION *023*

CONTEXT INHERITING
NANXUN PLANNING EXHIBITION HALL *026*
HUNNAN CAMPUS LIBRARY OF NORTHEASTERN UNIVERSITY... *038*
KUNSHAN WESTERN MEDICAL CENTER, PHASE 1 .... *048*
DUNHUANG PUBLIC CULTURE COMPREHENSIVE SERVICE CENTER.. *060*
BEIJING−TIBET COMMUNICATION CENTER... *068*
THE RAILWAY BRIGADES MEMORIAL.. *076*
DATONG MUSEUM *086*
ABOUT SYMBOL *095*

HERITAGE CONSERVATION
MAIN PAVILION AND AUTOGRAPH COLLECTION IN JIANGSU GARDEN EXPO................................... *098*
JW MARRIOTT HOTEL QUFU *120*

URBAN RENEWAL
SCHOOL OF URBAN DESIGN, WUHAN UNIVERSITY *130*
CHONGQING PLANNING EXHIBITION HALL *142*
KUNSHAN GRAND THEATER... *152*
BAOJI CULTURE AND ART CENTRE.. *160*
RENOVATION OF ENCYCLOPEDIA OF CHINA PUBLISHING HOUSE *170*
ABOUT CITY I: DIMENSIONS AND PERSPECTIVES OF URBAN DESIGN *176*

ABOUT CITY II: TO CREATE CHARACTERISTICS FOR URBAN FEATURES *179*

LANDSCAPE INTEGRATION

RIZHAO SCIENCE MUSEUM *182*

RIZHAO SCIENCE AND CULTURE CENTER *196*

LANZHOU PLANNING EXHIBITION HALL *206*

CHONGLI CENTER *214*

LIFESTYLE ADVOCATION

MAIN BUILDING OF TIANJIN UNIVERSITY NEW CAMPUS *228*

SHAHE CAMPUS OF BEIJING UNIVERSITY OF POSTS AND TELECOMMUNICATIONS *240*

TSINGHUA SHENZHEN INTERNATIONAL GRADUATE SCHOOL, PHASE 1 .... *248*

SHENZHEN VANKE CLOUD CITY *256*

JINAN SHUNTONG BUILDING *262*

QINGDAO SCODA PEARL INTERNATIONAL EXPO CENTER *268*

ABOUT STYLE *277*

RURAL REVITALIZATION

KUNSHAN XIBANG VILLAGE KUN OPERA SCHOOL & RURAL WORKSTATION *280*

KUNSHAN JINXI ZHUJIADIAN BRICKYARD RECONSTRUCTION *290*

SMALL LAND OF PEACH BLOSSOMS IN KUNSHAN *298*

REVIEWS & INTERVIEWS

ARCHITECT CUI KAI AND HIS "LAND-BASED RATIONALISM" *310*

LAND-BASED RATIONALISTIC DESIGN METHODS:
    AN INTERVIEW WITH ACADEMICIAN CUI KAI *312*

GROWING ON THE LAND:
    CUI KAI'S "LAND-BASED RATIONALISM" ARCHITECTURAL
    PRACTICE IN DUNHUANG *318*

DIRECTION AND APPROACH:
    AN INTERVIEW ON PRACTICE OF LAND-BASED RATIONALISM *335*

EPILOGUE *349*
ANNOTATIONS *350*
CHRONOLAGY *356*

引子
——本土设计方法论初探

2009年我出版第一部《本土设计》时，提出了"本土设计"的理念，当时主要是对以往自己设计思路做一种回顾和反思，找到内在的规律和共性。另外也是针对21世纪初开始的建筑行业开放市场状况，当时国际设计师带着国际建筑时尚语汇大举进军中国市场，无论从文化还是美学上对中国当代建筑界造成了很大的冲击——中国建筑师们不仅无奈地在几乎所有的大型公建项目中败北，而且也让他们失去了建筑创作的文化自信，一片迷茫，批评和抱怨声此起彼伏。因此我提出"本土设计"的理念就是试图重新阐明建筑与土地的关系，确立不同的地域环境应该孕育不同的地域建筑特色的基本逻辑。

这个理念一经提出，曾引起同行的关注，虽然有很多人表示支持，认为这个想法很及时、很必要，但也有不少人持怀疑的态度，认为在开放的背景下，创作自由度大增，一提本土和地域，会不会又走到保守封闭的老路上去？我为此也作过解释：第一，不是封闭，本土设计强调的是建筑和土地的关系，而不是建筑师和土地的关系；第二，本土设计是希望创新、寻求特色，只不过这种创新和特色不是跟风赶时髦而来，而是从本地的文化和自然资源而来。

2013年，我又出版了《本土设计Ⅱ》，在文字部分更系统地阐述了自己的本土设计主张，并用一系列的原则和策略来指导实践，创作出了一批有鲜明地域特色的建筑作品。此书一出，就受到同行的欢迎，在社会上也有不少积极的反响，至今仍有不少甲方和政府领导向我索要这本书，而我办公室的存书已经不多了，所以从去年开始就在着手准备《本土设计Ⅲ》。

与前两本书出版的时代背景相比，这几年国家在政治、经济以及社会各个层面又进入了一个大转型的时期。党的十八大之后，党中央对之前经济快速发展中出现的许多普遍性问题和屡禁不止的社会顽疾"动了大手术"，严厉治理和督查，得到了人民的热烈拥护。对我们建设领域来说，党中央从生态保护大战略的"两山"理论到城乡特色保护的"乡愁论"，还有针对城市更新提出要下"绣花的功夫"，在建筑标志性设计中要杜绝"大洋怪"的要求……在不同层次、不同方向上都给出了十分专业可行的理论和实施路径，彻底扭转了城市建设中的粗放扩张、攀比高度、贪大求怪、拆古作假、推高地价房价、城市更新和治理成本不断加大、不可持续发展等一系列不良之风，让各级政府领导冷静下来，端正政绩观，重新用理性的视角去思考城乡发展的未来。

我发现，这几年我的本土设计思想越来越能在与领导和甲方的交流中产生共鸣，我们建立在本土设计理论和方法上的设计方案的通过率也越来越高；甚至项目之前别的团队提出过若干方案都未被认可，请我们上手就一次通过，用个不大恰当的比喻，真有一种"药到病除"的感觉。这其中"病"就是要找准设计的问题所在，不盲目；"药"就是以本土设计的相关策略和方法去解决问题，"对症下药"，不瞎做。这种方式我称为精准设计，往往只出一个方案就行，只要它有说服力，只要业主领导有共同的价值观，方案就会一次通过，效率很高。因此这几年我们团队创作的作品数量越来越多，品质也在不断提升，本土设计的理论在业界也得到越来越多的认同。

不少同行建议我将这种设计策略再系统化，总结出一些可以传授的解题思路和模板与大家分享，这让我的团队备受鼓舞，也有了进一步总结、提炼和反思的动力。2021年秋季，北京建筑大学建筑与城规学院邀请我和团队的主要助手骨干一起给研究生讲授"本土设计导论"，这便成了本土设计方法和模式语言研究的真正机缘。当然，这个总结和研究还需要一个过程，从讲课的PPT到讲义到教学参考书，对我们忙碌的设计师来说都是个艰难的任务，需要许多专家学者的支持、指导和帮助，它在《本土设计Ⅲ》中还只是初步的、片段的呈现。

在本书中，作品被分为"文脉传承""遗产保护""城市更新""风景融入""生活引领""乡村振兴"六大类，与同期出版的《本土设计Ⅳ》中的"绿色导向"类作品一起，共同组成了"本土设计方法"中的七大方向。

崔愷
2022年8月

INTRODUCTION:
PRIMARY EXPLORATION OF LAND-BASED RATIONALISM METHODOLOGY

When I published my first book *Native Design* in 2009, I proposed the concept of "Land-based Rationalism", which was a sort of reflection of my design philosophies, through which I could find inherent rules and commonalities in my projects. It was also a response to the opening-up of the architectural market in the early 21st century, when foreign designers were flooding into Chinese market, not only making a great impact on contemporary Chinese architecture both culturally and aesthetically, but also hurting Chinese architects' confidence because of frequent defeats in bidding projects, triggering widespread complaints and criticism. Therefore, I proposed the concept of "Land-based Rationalism" in an effort to re-explain the relationship between buildings and land, and establish a fundamental logic that suggests different environments nurture different regional architectural characteristics.

Once proposed, the concept attracted the attention from the industry. Although many people expressed their support, there were many skeptical voices, believing that mentioning "land-based" may lead us to the old path of conservatism in the background of opening-up. Firstly, "Land-based Rationalism" is not isolationism. Instead, it emphasizes the relationship between buildings and land rather than that between architects and land. Secondly, "Land-based Rationalism" seeks innovation and unique features, which are not derived from following trends, but from local culture and natural resources.

In 2013, I published Land-based Rationalism *II*, which explained "Land-based Rationalism" in a more systematic way. It showed architectural practices with regional characteristics, which were all implemented with a series of principles and strategies of "Land-based Rationalism". Upon its publication, it was welcomed by my peers and received many positive responses from the public. Until now, many clients and government leaders are still asking me for a copy of Land-based Rationalism *II*. So I have been preparing for Land-based Rationalism *III* since last year.

Compared with the background when those two books were published, in recent years, our country has entered a period undergoing great transformation in many aspects of the society. Since the 18th National Congress of the Communist Party of China, The CPC Central Committee has carried out a major campaign to deal with many common and difficult problems that arose from rapid economic development in the past, and has gained great support from the people. In the building industry, The CPC Central Committee has made many clear instructions, providing professional theories and implementation paths at different levels and directions, thoroughly reversing a series of inappropriate trends in urban construction. He encouraged governments at all levels to look at the future of urban and rural development from a rational perspective.

In recent years, I have found more common ground with the authorities and clients in "Land-based Rationalism", and our design based on "Land-based Rationalism" have become increasingly successful in getting approval, as if it is a good medicine that cures "illness" in design precisely and timely. The design always gets approved effectively without too much modifications as long as we share common values with the clients. Therefore, the quantity and quality of our projects have been continuously improving, and "Land-based Rationalism" has gained more recognition from the industry.

Many peers suggested that I further systematize my design philosophies and summarize some problem-solving ideas that can be shared. This has inspired my team a lot. In the fall of 2021, the School of Architecture and Urban Planning at Beijing University of Civil Engineering and Architecture invited me and backbone members of my team to give the course of "Introduction to Land-based Rationalism" for postgraduate students, giving us the opportunity for researching our "Land-based Rationalism" methodology and pattern language. It's not easy to complete such a task in a short period of time, and what we present in this book is just a preliminary and fragmentary part of our research.

In this book, the works are divided into 6 categories: context inheriting, heritage conservation, urban renewal, landscape integration, lifestyle advocation, and rural revitalization. Together with the works of "green orientation" in Land-based Rationalism *IV* published at the same time, they constitute the 7 directions of " Land-based Rationalism Approaches".

CUI Kai
August, 2022

设计方法

DESIGN METHOD

关于方法　ABOUT METHOD

　　设计要有方法。

　　以前学校里教的多是如何构成空间和搭建造型的方法，关注点在建筑本体，是由内而外的。而本土设计的重点是处理建筑和所处外在环境之间的关系，是由外而内的。这并不是先外形、后内部设计的意思，而是先分析外部环境、再寻找自身应对策略的逻辑，因此设计方法就不一样了。当然学校里学的方法仍然有用，是基本功，没了基本功，建筑本体构建不顺，再加上外部条件的介入，就会乱了章法。许多比较差的方案往往有东拉西扯的拼凑之感，一看就是设计的基本能力问题。所以讲本土设计应该是建立在比较扎实的基本功之上的。

　　本土设计方法主要是来自我主持的本土设计研究中心大量工程实践的归纳和提炼，也从对国际大师作品的考察学习中发现了类似的规律：比如在历史街区中的项目通常都采用高度一致、尺度协调、色彩或材质延用等方法；在自然景区中也常见以消隐体量、建筑与景观一体化，以及创造看风景的场所和视角等方法来建立建筑与环境景观的关系；面对城市生活，都有开放空间、强化场所感、促进交流共享的策略；面对生态绿色的主题，也有从设计到技术的一套体系，只不过在发达国家更强调主动式技术，而我们根据国情强调的是被动优先，主动优化，充分调动建筑设计的方法，从自然通风、采光、灰空间利用等入手，让建筑减少用能时间和缩小用能空间，让绿色建筑有地域的气候特色。当然许多工程面临的问题是综合性的、多向度的，本土设计的设计策略也是多向度的，提出综合性的解决方案。这听起来挺复杂，但其实每个项目都会结合自身的问题，有侧重地找出设计目标和权重，而并非一一对标，面面俱到。

　　因此，设计之初，通过踏勘和调研，先找出问题是关键的环节，然后经过讨论和思考提出基于本土设计价值观的目标和设计策略，在这些都比较明确的情况下，再集思广益，找到最适宜、最有效的方案。之后在初步方案的基础上，安排建筑功能，提炼设计语言，选择技术手段，结合项目的投资控制和工期限制，不断深化方案，形成最优的成果。

　　这一套理性方法和价值观引导下的设计是希望提高设计解决问题的精准度，少走弯路、错路，少折腾，将思考聚焦，将建筑的形态和设计语言的选用与解决问题的目标、方向契合。如此引导设计的方法基于这样一种具有普适性的假设：路子对了，设计好一点就是一种理性的创新；路子对了，设计差一点也不是大问题；路子错了，设计再好也不能解决问题，仅仅成为建筑师自我主观的表达；路子错了，设计再差一点，就是一个既主观又没品质的设计。显然我以为本土设计的控制点是追求第一类，宽容第二类，避免第三类，拒绝第四类。为此，我和团队合作，试图从以往的有代表性的实践案例中总结和提炼出一系列的模式语言，将面对的问题、解决的策略、设计的方法一一对应起来，在场地信息的分类框架下找到对应的结合点，形成一种有逻辑的设计规律，以供初步掌握了设计基本功和相关知识的同学学习，也适合刚入行的青年同行们了解与参考。

　　当然，建筑创作是一种影响因素很多、决策判断比较个人化和被动式的设计活动，很难像数学解题似的严密逻辑，更不会只有唯一解的情况。但对量大面广的建筑来说，综合解决问题仍然是一种基本的目标和价值底线，因此从这个角度来看，设计是有对错之分的，是有标准、有规律可循的，所以对设计的方法研究也一直是设计界和教育界反复研究的基本问题。我们以本土设计的价值观为线索梳理出来的设计模式语言也是建立在以往研究的基础之上，许多原理和常识无疑是一样的，但又有自己的侧重点和方向性，希望对当下我国建筑设计有一定的应用价值和现实意义。

本土设计方法思维导图　　MIND MAPPING OF LAND-BASED RATIONALISM DESIGN APPROACH

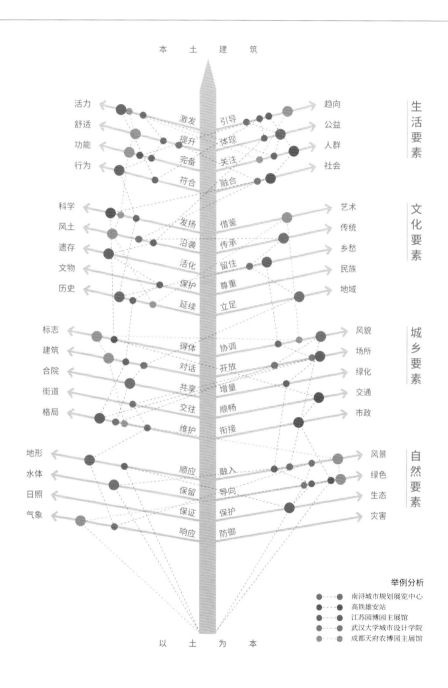

本土设计

是一种开放性的设计思想

同时也是一系列方法和策略的组合

方 法		
	调研踏勘	调研是设计的起点，对项目现场和任务的全方位调查和研究、对相关案例的学习借鉴十分必要，以往个人观察和经验的积累也很重要
	分析解题	对调研信息进行梳理、解析，找到关键问题，找准解题策略。这是一个试错的过程，在策略方向上比较多种空间模式和技术路线的可行性和优缺点，筛选出最佳路径
	明确立场	价值观十分重要，它在每一个选择点上是贯彻始终的判断标准。如果没有明确的价值观，判断选择就会失去依据，造成犹豫不决和主观盲目
	寻找平衡	平衡点是各种观点和权益交流碰撞的结果，主动、尽早找到平衡点是决策的基础。要主动了解各方诉求，尊重各方利益，认真倾听，积极回应，开放讨论，寻求共识
	因地制宜	因地制宜不是机会主义，而是客观性决定主观性的内在逻辑。设计不事前预设答案，在对场地和相关要素的逻辑分析中推演方案，排除试错后找到最佳方案
	将错就错	不是宽容和将就，而是不放弃、在既有条件下争取持续控制力的积极态度。避免被屏蔽，保持良好的合作沟通氛围，让工程全过程处于有效的设计控制之中
	中庸之道	这是一种处人处事的灵活性和平衡度，是解决各种矛盾的开放心态。基于建筑是社会产品的基本认知，相信人人会理解彼此、事事变化有内在逻辑，不急躁武断，不固执己见，在探讨和调整中坚持初心

方向	目标	策略
文脉传承	大文化视角，从多层次多类型的地域文化载体中汲取营养，与当代建筑的空间、结构、当下生活相结合	借形、格局、肌理、符号 材料、色彩、意境、隐喻 艺术、生活
遗产保护	保护历史遗产信息的原真性，尽量不改变历史遗产已存在的环境特征，营造适合展示历史遗产文化信息的空间场所	消隐、控高、保护、展示 对比、对话、协调、原真 利用、修缮、风貌
城市更新	城市是有生命、不断进化的综合体，城市更新是织补，也是演进，要保护和利用并重，修补和创新并举，旧中出新，新中有旧，要顺应和描述城市进化的状态	协调、织补、重构、演变 渐进、将错就错
风景融入	敬畏、保护自然，让建筑融入风景，作为景观环境的有机组成部分，并为人们创造体验自然，观赏风景的空间场所	覆土、嵌入、顺形、借势 隐退、靠色、轻透、映影
绿色导向	少扩张多省地，节省土地资源；少人工多天然，适宜技术的应用；少装饰多生态，引导健康的生活方式；少拆除多利用，延长建筑的使用寿命	遮阳、导风、保温、隔热 开放、除黑、采光、灰空间 绿植、再生、减排、蒸发、集约
生活引领	关注社会生活，跟进需求变化，创造人们喜爱的场所，促进公共空间的共享、公平和善意，是本土设计以人为本的基本策略	共享、分享、共建、混搭 关爱、公共性
乡村振兴	守护田园，留住乡愁，要因地制宜、轻设计、陪伴式，保持和提升乡村特色，助力产业发展，帮扶弱势群体，推动乡村可持续发展	针灸式、保格局、留乡愁

策略

关于场地踏勘　ABOUT SITE SURVEY

　　本土设计强调建筑创作源自对场地的认知，所以对现场的踏勘是不可忽视的设计起点。

　　无论多忙，每一个新项目，我都尽量争取和助手一起去看现场，尤其是场地环境有特色、地形地貌复杂或周边限制条件苛刻的地方，我特别有兴趣去现场走一走，不看不踏实。以前的确出过这种情况，助手去过现场，说就是一片空地，没什么要保留的，所以就按规划红线和任务书开始设计。而当我参加方案讨论时，在卫星图上看到了一列大树，便问道："这些树在建筑布局中考虑了吗？"助手说这不是规定要保留的树，只是农田水渠边的大杨树。我一下来了精神，马上安排时间去现场补完踏勘，结果发现这行呈十字形的大杨树在宽阔的田园中赫然而立，很有标识性，这些树的走向顺着农田的田垄沟渠，与规划网络还偏转了十几度。我马上决定以树阵为核心重新布局，把原本一个体量切分为几块，以树渠为轴形成新校园的步行轴。这样不仅大树保下来了，还使建筑的场所感大大提升，好像把建筑"锁定"在大地上似的。

　　还有一次，有个南方的小城市请我去做几个文化建筑，事先发来了上位规划图，场地位于一个圆形广场周边，呈围合的状态，没有介绍什么令人兴奋的场地信息。不久后，我在出差行程中改道去看现场，赫然发现有一座郁郁葱葱的小山，便问这个小山在规划图中的什么位置？答曰："就是这个圆形广场哦！"那里已经有一台推土机在挖山，没几天就将被推平了！我当时就急眼了，马上和陪同的领导说："能不能先把土方工程停下来，如果这个项目是以削山为前提，我宁可不做了！如果山能留，我们可以帮忙修改规划，让城市广场和文化馆群落与山水更好地融合起来。"领导当即表态可以调整，哪怕调规要花点时间也可以。于是我当晚在酒店就勾出了草图，把山和馆用环形屋顶平台连成一体，不仅山留下了，也让这个原本设计平庸的圆形广场格局变得很有特色。虽然因为投资原因这个项目进展缓慢，但能把山保下来，就是做了件善事，心里特别踏实。

　　其实对我而言，现场踏勘并不仅仅是了解环境，拍些照片，回去慢慢分析再开始构思方案，而往往是带着设计任务看现场，边看边想，把眼前的现场情况和脑中的任务功能叠合在一起，判断出在哪儿摆什么功能合适，在什么功能空间应该看到什么场景。这种在头脑中的信息对接和图像叠合，往往就产生了方案的构思，在回程的路上就很快勾出草图交给团队，带有一种强烈的现场感和新鲜度，一般这样产生的方案成功率都很高。

　　比如武汉大学基建处请我去设计城市设计学院教学楼，听说之前学院自己做过许多内部方案比选，学校都不太满意，刚好我那之前在珞珈讲堂做过一次学术报告，学校就有意请我来做。我特别喜欢武大校园中的历史建筑群，尤其是巧妙地利用山地条件，把学生宿舍放在轴线上形成台地建筑，我当年读研期间去参观时就留下了深刻的印象，但近几十年的校园就不太讲究了，也没有传承武大的文脉，所以我想在城市设计学院的设计中能否把这条线索续上。新学院楼就在原址上，之前只是利用武汉水利电力学院附中的教学楼改造而用，用地虽小，但就在珞珈山脚，位置重要。我和助手头天晚上入住珞珈山庄后就沿山中绿道绕了两圈，第二天一早又在饭前跑到山上去看那些经典历史建筑，一直走到现场，在当时未拆除的楼上楼下走了一遍，边看边想在这片狭长的山脚用地上如何让设计教学空间与山地景观协调，还能形成交流互动的场所，如何利用屋顶让眺望东湖的景观平台与教学模型的实体搭建结合起来？以及如何把武大历史建筑厚重感的构筑方式与现代材料和结构体系结合起来？一路走走停停、看看想想，转完现场我心中就有数了，头脑也兴奋起来，很快就把草图勾勒出来，设计的方向也随之确定。后来方案得到了校方的认可，学院老师们也颇有惊喜，在这条用地上也可以营造出开放的"创意社区"！

　　近几年来，我们本土设计研究中心的方案创作成功率比较高，能解决问题比较精准，主要得益于对场地的踏勘、分析比较到位，而带着任务看场地又较之以往"盲目看"更有目的，也能马上把现场观察到的信息与任务要求衔接起来，使方案大构思在现场完成，这的确是本土设计的一个重要创作方法。

关于判断力　ABOUT JUDGEMENT

设计是不断发现问题和解决问题的过程，而对解决问题路径的选择则是要不断地作出判断。

判断得准而快，会使构思顺利推进，从一个问题的解决转到另一个问题的解决，抑或作出一个好的判断能使一系列相关问题迎刃而解。但如果判断错了，显然会导向失误，不仅不能很好地解决问题，可能还会带来更多的问题，将设计引向一种盲目或放任的状态，失去了理性的逻辑。如若判断不清、总是犹豫不决，不仅会拖长进度，更容易让构思过程不断反复，打乱构思的整体性和连贯性，而且耗费大量精力，做得很累，也无法有力地、精准地解决问题，往往采用多方案排列组合、拼凑的办法，显然也出不来好想法、好方案。

因此，要想提高设计的效率和创作构思的精准度，提高判断力是必须的。判断力也分不同的层级：一种是带有价值观的判断力，比如对本土性的价值观、对绿色生态的价值观、对历史遗产保护和利用的价值观、对社会性的价值观等——这个层次的价值观判断可以有对错之分，是一种伦理的判断；再一层级的判断力是和建筑学本体有关的，比如关于形式美、关于空间构成、关于结构逻辑、关于材料系统等，都是建筑师基于职业素养而形成的判断力——这种判断力既有职业公认的标准，也有个人经验和个性化的成分，在设计中这个层级的一系列判断是分辨设计水平高低、趣味好坏的依据，也是我们常说的有没有设计感。除了这两个层级的基本判断力之外，还有一些判断力对掌控设计的实施十分重要：如对市场和形势的判断力，对于承接项目、拓展服务以及回避风险都很重要；如对经济成本的判断力，对控制造价、保证质量、顺利推动项目的实施至关重要；如对施工工法和材料选择的判断力，对管控施工质量、应对选材变化、把控最终效果十分关键。

判断力对设计如此重要，那么如何提高自己的判断力呢？如何能在复杂综合的工程项目中作出快速和正确的判断呢？只有一条路，就是不断学习和积累，要不断地锻炼，高强度地锻炼。

以我自己的经验而言，对判断力的培养也经过了相当长的过程，从大学时代，我就比较注意老师指导和评图的方式方法和点评意见，比如人的使用行为和空间设计的关系，比如色彩和材质、与立面空间构成的关系等。老师的指点也使我养成观察建筑使用状态的习惯，直到现在我在观察现场、观察城市中都还是延续这种处处观察、反思、推断的方式；看图时，也往往很快能发现违反使用行为的功能布局和空间设计上的毛病，不厌其烦地进行改图纠正。对方案创作方向判断力的培养更多地来自看别人的设计：从上学时爱看邻桌同学的设计和老师的改图；到工作中爱看同事的设计项目，有时还愿意瞎出出主意，人家听不听也无所谓；到后来参加竞赛评审会，从那些设计方案中的确能学到很多思路、手法以及阐述的方式和技巧，当然看到比较差的方案也能比较清楚问题所在，有时还提些修改优化的建议。这些对锻炼自己的判断力都帮助很大。其实还有一种十分重要的提升判断力的方法，是向业主和领导学习。他们虽然一般不是专业人士，但从我们所不太熟悉或关注的视角看设计是他们的长项，比如领导常常从城市发展、规划格局、现实困难和可操作性方面考虑较多，业主往往从投资回报、利益最大化等方面考虑。不少建筑师都爱抱怨领导的干扰和业主的唯利是图，但冷静下来的话，我们应该承认他们观点中那些理性的思考和合理的诉求是设计应该考虑和努力满足的。而如果在创作构思中主动从他们的角度思考，沟通就容易得多，方案就很容易被接受，未来的使用也会更合理。实际上这些年业主团队中也不乏有设计背景的技术管理人员，与他们沟通可能效率更高，因为把甲方的一些经验性想法"转译"成技术要求就更容易被理解和达成共识。

不过近几年，我越来越清楚地认识到，要提高和达到精准的判断力，设计价值观是至关重要的。以往创作型建筑师一般比较在意建筑美，从经典的到流行的，甚至是个性的建筑形式语言的使用或创造是大家很在意的，甚至一旦被否定就很沮丧，一旦没有形式创新的机会就觉得设计没什么意义，也没什么价值了。这种观点虽然从艺术的角度是有意义的、可以理解的，但面对建筑所包含的综合性价值来说就显得很偏颇。实际中持这种观点的建筑师会刻意夸张形式的价值而不惜牺牲其他的价值，如功能和经济性，而对形式过度关注的设计，也往往会忽略或伤害许多基本的功能。形式夸张的建筑往往出现不好用、太浪费的问题。而更多工程型的建筑师熟悉各种规范和条文，他们的设计价值观主要是让工程合乎标准和规定，因为这些规范、标准的确包含了行业专家的智慧，代表了建筑达到相关使用要求的底线，当然是在设计中作出判断的依据，但总体上来讲，这类判断是相对比较被动的，很难在创作中产生向前的推动力和指导性，而事实上这类建筑师一般是站在创作型建筑师的后面发挥技术支持和设计优化的作用，并不主导设计的总体走向，对方案做判断式的讨论参加不多，锻炼的机会少，在工作中属于实干而低调的一个群体，也令人尊敬。我在此描述这两类建筑师各自在设计中的状态和判断的层级，实际上并非是为了说孰轻孰重，而是想说，有没有一种价值观是能够跨越前后两阶段、对设计的每一步走向作出一以贯之的判断呢？我发现是有的，这就是绿色建筑的价值观。

关于绿色建筑的价值理念我以前也说过，刚开始是"加法"：为了节能要给建筑加上许多设施设备，评价标准是按项打分，加一项就加几分；还会增加投资造价，业主不愿意加是考虑造价，建筑师不情愿加是担心影响美观。而近几年思路出现了转变，在建筑师参与主导的绿色建筑设计中主要是先用"减法"，从减少建设用地、减少拆房、减少侵占自然，到减体量、减用能空间和时间、减少玻璃幕墙，乃至减轻结构、减少装饰、减运行成本，这一个"减"字，从规划管到建筑、管到结构和设备，也管到室内、管到景观，成为一个系统性的判断方法。因此只要有节能减排、创造绿色建筑的初心，只要有节俭生活的体验和常识，就可以对设计作出一系列基本判断：建筑布局是不是有利于集约用地、保护环境？既有建筑是不是可以不拆而加以利用？建筑形态是不是符合场地的气候特点？内部空间是不是大小适度、利于采光通风？结构造型是不是可以更轻量化，更利于灵活使用，更适合露明？设备方案是不是更高效、更灵活、更适度？室内设计是不是以少装修、少伪装为目标？景观是否更自然生态而不是刻意装饰化、人工化？凡此种种，都是在设计不同阶段作为判断的价值标准。满足了这些价值标准之后，才是美的问题、形式的问题、文化的问题。所以我想应该倡导一种绿色建筑的新美学：它是绿色的，也一定是地域的；它是现代的，也一定会反映传统的智慧；它应该更通透、更轻巧、更集约、更真实、更质朴。也许，它能代表中国建筑乃至世界建筑的未来。

近几年来，我自己的创作过程总以绿色建筑的价值观作出判断，校正方向；在指导团队设计时也是以此引导和评判；在设计深化中，甚至在建筑实施过程中对选择材质、选定工法以及在指导室内和景观设计中，都用同样的思路去作出选择，正所谓"一以贯之"，毫不犹豫。前不久，我主编的我们集团的《绿色建筑设计导则》正式出版发布了，我想它的作用就是给建筑师，也包括工程师，在设计中有一个作出正确判断的指引。如若大家在各个阶段、各个环节都能以"绿色、节能、减排"这个价值观去作出正确的选择，那建筑就不可能不绿，城市就不可能不好，自然就不可能再被粗暴地伤害，这是我当下认为最重要的事情。

关于沟通　ABOUT COMMUNICATION

建筑作为一种公共产品，它的设计就不是个人化的"自说自话"，也不是设计者个人情绪或理念的单一表达。设计需要沟通，设计的过程就是沟通的过程；设计的结果也就是经过沟通所呈现出的各方意向达到平衡或共识的结果。这是建筑区别于艺术的一种属性，也是建筑师区别于艺术家的一个原因。

对我来说，沟通第一个作用是学习，学习对方提供、你所不知道的信息或知识。几乎每一个新的项目都需要从甲方、规划局、机关策划团队、相关专业设计团队等各方面了解新的信息、学习新的知识。这些信息对理解设计的目标、任务都十分必要，掌握得越多越深入，你对设计的判断就越准确，方案就越靠谱，就越容易达成共识。反之，如果不主动去沟通、不积极去倾听、不认真地去学习，设计往往会主观和盲目。有时候哪怕忽略了一个看起来并不太重要的信息，到后面都有可能造成方案的彻底翻盘。这种教训我以前碰到过不少次，教训深刻。其实，对沟通中的学习也不必特别功利，似乎只为了做这个项目，事实上任何学习所积累的知识或见识都会成为有用的储备，也是积累知识、丰富阅历的重要渠道。坦率说，我们今天所需要的知识没有多少是读大学掌握的，之所以我们今天能够驾驭设计、受到业主和各方面的尊重和信任，绝大部分是在工作中不断学习和积累的。因此，如果以学习的心态去与各方沟通，你就会放下"身架"认真面对，就会对对方多一分尊重、多一分感激，也就会认真地记笔记而不是心不在焉地旁听。许多年轻的建筑师往往因此而错过了学习的机会。

沟通第二个作用是交流。交流是不同观点的碰撞和分享，交流的结果可能达成共识，也可能不能达成共识，但通过交流至少双方或多或少了解了对方的观点，这也是一种有价值的结果。交流的另一要点是倾听，要专注观察对方的话语内容和说话时的表情，这些都会有助于读懂和了解对方的态度和观点。对对方最大的不尊重往往出自"走神儿"，完全不在意或没听见对方说了什么，这很容易把事情搞砸，使交流中断，进而影响双方合作的关系，这是正式工作交流中很愚蠢的、不该出现失误。交流中除了要听懂和了解对方的看法，更重要的是及时发现或抓住有价值的核心内容和立场，尤其是要判断出哪些是要尊重的、哪些是与自己有共识的、哪些有误解要进一步解释的。听懂了，有了较清晰的判断，在对方讲完之后才能发言去进一步交流沟通，这时候也要注意对方的表情，察言观色，发现哪些对方感兴趣、哪些对方不太在意、哪些对方同意你的观点、哪些面露难色肯定不同意。根据这些表情你需要随时调整自己的发言内容和节奏，该讲的讲到位，不该讲的不要楞讲，至少可以想办法换个方式讲。在设计方案交流中，最重要的是要争取甲方的认同点，认同的点越多，方案成功的把握就越大。一般这些观点由会议室中不同的人表达出来，但关键还看最高决策者的态度，这中间变数很大。但无论怎样，通过有效的交流，找准问题所在、找到解决的方向就算达到了目的，毕竟建筑方案是非常复杂的综合信息的集成，很难一次精准地解决所有问题、得到全面认可，有五成以上的认同和赞许就算成功了。

沟通第三个作用是传播。设计是有价值取向的一种创作，在沟通中传播价值观是十分重要的。当今方案汇报都要做PPT，文件中前面几页都是一些纲领性的理念，但是这种标签式的理念并不能真正打动人，进而起到传播的作用。而在"面对面"的沟通中，将积极的、充满正能量的价值观用平和的话语、务实的方式、真诚的态度表达出来才是更有说服力的。在当今城镇化发展转型过程中，许多原来的理念都面临着纠错或转向，传播生态绿色、有机更新、活力创新的价值观是设计实践中十分关键的新的评价体系。因此，少喊空泛的口号，以务实的态度、引领性的视角去介绍理念、传播正确的价值观是沟通中的重要一环。

借形：将建筑的造型与传统建筑的典型形态相接近，以建立新旧之间视觉联想的关系。
格局：将传统建筑的布局形式或空间结构和场所特色用于建筑设计中，以达到空间感知的相似度。
肌理：将传统建筑重复性要素的节奏、尺度和材质纳入到设计中，以唤起视觉、触觉等感官的记忆。
符号：在设计中适当选用传统建筑中带有文化信息的符号，如纹样、雕饰、吉祥标识等，以传递相关的文化信息。
材料：传统建筑材料的利用或重构，特别是有历史信息的旧材料的保留和利用，以留下时间的痕迹。
色彩：设计中，对传统建筑的色彩基调的引用是强化新旧建筑视觉联系、传承本地文脉的基本手段。
意境：将传统建筑的精神意境在设计中运用和呈现，是在更高层次上进行文脉传承和表达的策略。
隐喻：以有典型特色的传统文化标识物为蓝本，在建筑形态、语汇或材料中加以适度表达，唤起公众的共鸣和联想。
艺术：多彩的传统艺术对设计会有多方面启发，将其渗透于建筑的形态、色彩、细部、空间陈设中，都能起到点题的作用。
生活：传统建筑中有特色的生活方式和行为可以在设计中作出响应，创造有针对性的空间环境。

文脉传承
CONTEXT INHERITING

本土设计主张大文化的视角，不再局限于民族形式和风格的传承，要从各个具体的地域文化中寻找文化的线索和基因，也要从多层次、多类型的文化载体中汲取营养，并非只是关注传统建筑和历史街区，虽然在实践中这仍是主要关注的文脉资源。最重要的是这些文化线索和基因要与当代建筑的空间、结构以及它们所承载的当下生活相结合，不应只是装饰语言，主要的是协调、呼应和对话。最终希望创作出有当地文化特色的、被当地人所认同的一种新的建筑语汇。

乡韵重弹
New creation of the traditional charm

南浔城市规划展览馆 · **NANXUN PLANNING EXHIBITION HALL**
设计 Design 2017 · 竣工 Completion 2019

地点：浙江南浔 · 用地面积：20 033平方米 · 建筑面积：9 180平方米
Location : Nanxun, Zhejiang · Site Area : 20,033m² · Floor Area : 9,180m²

合作建筑师：邢野、高凡、周益琳、曹洋、张笑彧、毛影竹、杨俊宸、洛丽贤
Cooperative Architects : XING Ye, GAO Fan, ZHOU Yilin, CAO Yang, ZHANG Xiaoyu, MAO Yingzhu, YANG Junchen, LUO Lixian

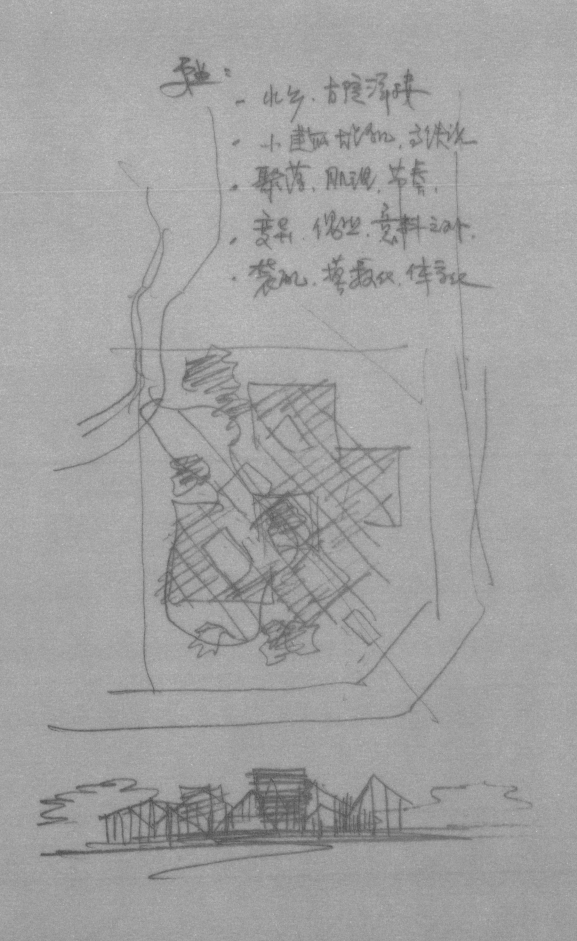

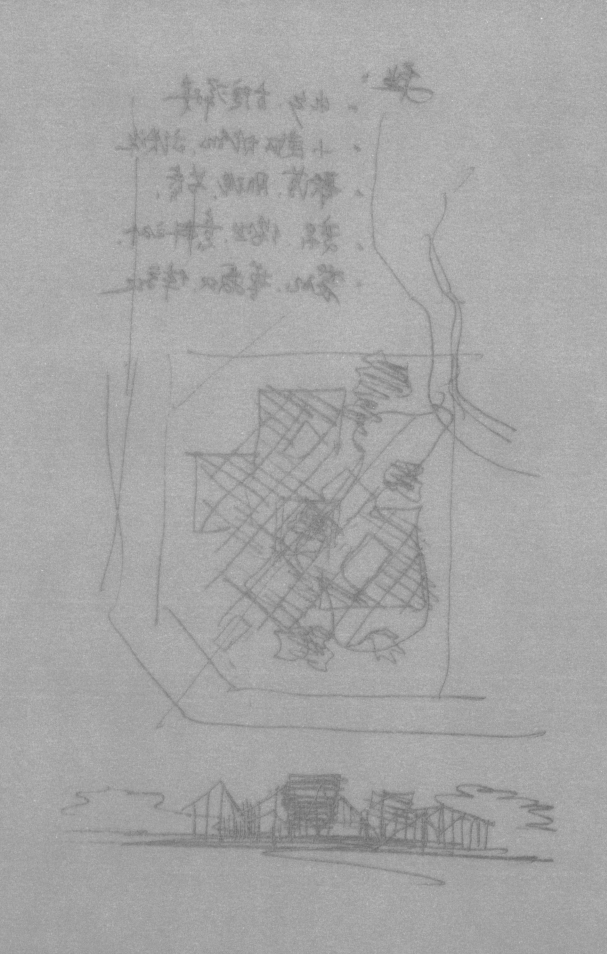

　　项目位于产业新城，北靠南浔古镇，基地周边水网密布。这里曾是江南典型的田园景观和水乡地貌，但随城市的快速发展，现状是被城市路网包围的荒芜空地。地处独特的地域环境，设计在传统与现代、自然与建筑之间寻找平衡点：一方面秉持现代语言，建立清晰有力的内在逻辑和自我展开的几何体系；一方面在类型学意义上向地方的空间传统学习，充分融合南浔古镇的坡顶聚落、园林庭院、骑楼水街、西洋砖楼等原型要素，最终旨在通过系统性的方法对地方传统空间语言进行更为内生性、结构性的现代转译和创新。

　　建筑以基本的"结构—空间"单元为起点，形成富有秩序感的聚落形态，仿佛绵延水乡肌理的截取片段。空间组织汲取南浔特有的宅第院落内嵌西洋砖楼模式，将西式砖筒作为核心空间嵌入外围钢屋架系统。底层架空区域形成开放的公共城市空间，曲折的水街穿越其间，为市民提供漫步休憩的园林空间，营造"园中有房，房中亦有园"的场所意味。

　　采用钢框架结构，通过外露钢架的方式将形式与结构的同一性充分表达。追随江南粉墙黛瓦、木作门窗等传统意象，立面上深灰色结构之间填充白色穿孔埃特板，内庭院界面采用竹木墙板、格栅幕墙。

The project is located in the new industrial area of Nanxun District, adjacent to the famous ancient water town. Located in such a unique geographical environment, architectural design needs to find a balance between tradition and modernity, nature and man-made. On one hand, the design uses modern language to establish a clear and powerful internal logic and self-expanding geometric system. On the other hand, it learns from the local spatial tradition in the sense of typology, fully integrating the sloping roof settlement, traditional Chinese garden, arcades, water streets, and western brick buildings. Overall, this design strategy leads to a more endogenous and structural modern reinterpretation and innovation of the local traditional space.

As a result, the building is based on a spatial structure unit. By fully absorbing the unique pattern of traditional house courtyards, the spatial structure units are systematically organized around four Western-style brick cores which embedded in the continuous steel roof truss system. This specific orderly settlement is just like an excerpt of the spreading texture of old water town.

The identity of the building is fully expressed through the exposure of steel frame. Following the local building images, such as white walls and black tiles, wood doors and windows, white perforated panels are filled between the steel structures on the façade, and bamboo and wood wall panels and grille curtain walls are used in the inner courtyard interface.

借形：借传统建筑之形，用当代装配技术之术，求文脉传承延续之新。

色彩：建筑之色因时而变，时间的痕迹在用色时要有预判并加以利用。

肌理：青瓦白墙交织而成江南水乡的肌理，顺其重构可得"和而不同"的效果。

格局：水在巷中，房在水边，典型的水乡空间格局应该是营建场所的基调。

   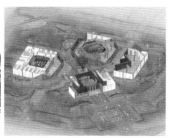

意境：水、石、树、径是庭院造景要素，也是水乡园林之意境所在。

生活：水乡的生活在水边，无论过去还是现在，生活的特色不能变。

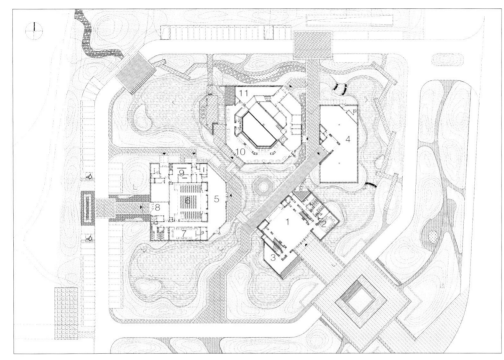

1. 主展厅入口大厅
2. 接待
3. 纪念品商店
4. 民俗文化展厅
5. 报告厅前厅
6. 报告厅
7. 会议
8. 接待区门厅
9. 员工食堂
10. 咖啡厅
11. 机房

首层平面图

剖面图

南浔城市规划展览馆·2017—2019·NANXUN PLANNING EXHIBITION HALL

设计随笔

南浔城市规划展览馆不仅是城市中的建筑，也是风景中的建筑，前面是车水马龙的湖浔大道，背后则是南林湿地生态公园。城市环境要求建筑具有一定的体量，要具有公共尺度的开口和室外共享空间，风景则要求建筑尽可能融于环境，"风景中的建筑不仅为人提供服务，亦不可破坏风景，最好是融入风景，或成为风景的一部分，甚至自成一景，为风景添上一笔，或成为风景中的焦点"。

在设计初期，我们就意识到设计面对的是城市和自然的双重叙事。设计草图一方面近似现场写生一样描绘出一幅水乡聚落的原风景，地平线上的坡屋顶连绵起伏，这可以说是站在现场面对平坦的水乡平原，会产生的最自然的想象，它源于对乡村地景的一种原始记忆；但另一方面又意识到和城市的尺度关系，建筑则集中高起，上部屋顶轮廓漂浮于绿树之上，希望从未来高铁站隐约可见，下部水系蜿蜒进入，架起仿佛水街骑楼一样的城市空间。因此，双重叙事之下，我们要做的肯定不是对地方风土建筑的简单再造。记忆需要延续，但它更应该是一次崭新的公共性营建，也是一次尝试将地方传统语汇进行现代转译的机会，这正是Native Design 和Land-based Rationalism先后两个"本土设计"译词之间的根本区别。语言实验对于一个旨在展示水乡过去与未来的公共文化建筑而言，无疑是题中应有之义。与传统房屋相比，展馆的尺度、跨度、集中程度和开放程度，都已发生了明显的转移和变化，必然需要寻找相应的具有内生性的形式表达，而不是对传统形式在修辞层面的套用。

我们的切入点在于以一个几何性的深层结构将形式、空间、结构整合起来，将坡顶单元组合的聚落形态和底部斜撑架起的水巷公共空间结合在一个组构性的体系内，一切丰富性包含于一套几何逻辑之中，以一个体系最大化地将复杂的经验和理性统一起来。

首先，正、斜两套网格相互转换。坡顶单元按照8.1mX8.1m的基础网格紧密排布，而正交网格的对角线构成了另一套大约11.45mX11.45m的斜交网格。斜交网格的出现是整个建筑的机锋所在，它决定了几个重要的事实：1）斜网格定义了斜向屋脊，外围正向立面与斜向坡顶结合，一方面以最简原则实现传统聚落繁复生动、层层错落的屋面形态，一方面立面呈现为对屋面垂向暗示出的体积的剖切效果，既肯定又意外；2）东南角的礼仪大厅采用东南45度的方位朝向，从屋面体系来看它是唯一一个具有正面性的朝向，符合对礼仪入口的设定；3）因为其上只有轻质屋面荷载而没有楼层荷载，二层展厅空间可以相对一层8.1mX8.1m轴网减柱，转换为11.45mX11.45m轴网，结构尺度巧妙地发生转换，空间放大更有利于布展需求。

然后，平面中的斜交关系进一步向空间发展，转化成一个三维空间结构体系，这一方面形成了连续的屋架系统，几何秩序规定着其下空间；另一方面在立面和空间中出现体现传力路径的斜撑，以此塑造了内外一体化的格构立面和充满褶皱的巷、院空间。

另外，设计过程中，老宅中意外发现西洋楼的空间经验使我们念念不忘，颇具转折性的一个想法就是从想通西洋楼如何引入整体架构开始。四个独立分散的砖楼体量，它们依据几何系统生成，分别容纳相应的功能，并意象性地支撑起连续的钢构框架。随着设计的深入，四个砖楼因对应功能的精确调整和所处方位的具体特点逐步分化为形状和尺度各不相同的子类型，构成了"在语言学意义上那些语素诸线索的系列性"。砖楼为"型"，框架为"构"，"藏型入构"的做法与张石铭和刘氏梯号两座旧宅异曲同工。

在南浔城市规划展览馆中，坡顶单元类型因为带有重复（自我指涉）和表意（指涉传统）的特点，显然是符号性的。符号上升为系统，以虚实嵌套的方式在各个立面反复出现，大部分是"实中带虚"，房屋一样的开口嵌入白墙当中，唯独东南入口是"虚中带实"，一个白色房屋在透空的门廊当中戏剧性地反转出现，提示着局部与整体的辩证关系，它可以引发多重解读，况且"符号的意义本身就是无限衍义的过程"。

符号化语言不同于被抽干意义的纯粹现代主义语言，甚至可以合理地将地方图式内化吸收于自身，这两者的区别，艾伦·柯洪在1970年代通过题为《形式与图像》的文章，从形式主义理论的角度对其深刻讨论过，他直接引用了克里斯托弗·雷恩（Christopher Wren）的观点——建筑中的"客观的美"仰赖几何学，然而所有其他的美则仰赖习俗。与此不无相关的是，今天有学人提出"风土现代"的三种实

践方向:"如画式再造""客观性建构"以及"场所感再生",它们被认为是从怀旧到反省的不同阶段,借用这个提法的有趣之处在于,对于这三种方向,南浔城市规划展览馆似乎在同时实践,但同时又与具体归类保持疏离,习俗与几何、图像与形式、风土与现代、怀旧与反省,在超越主义之争的今天我们希望有不一样的调和。注1

厚重的讲究
Dignity & delicateness

东北大学浑南校区图书馆 · HUNNAN CAMPUS LIBRARY OF NORTHEASTERN UNIVERSITY
设计 Design 2012 · 竣工 Completion 2018

地点：辽宁沈阳 · 用地面积：25 808平方米 · 建筑面积：43 700平方米
Location : Shenyang, Liaoning · Site Area : 25,808m² · Floor Area : 43,700m²

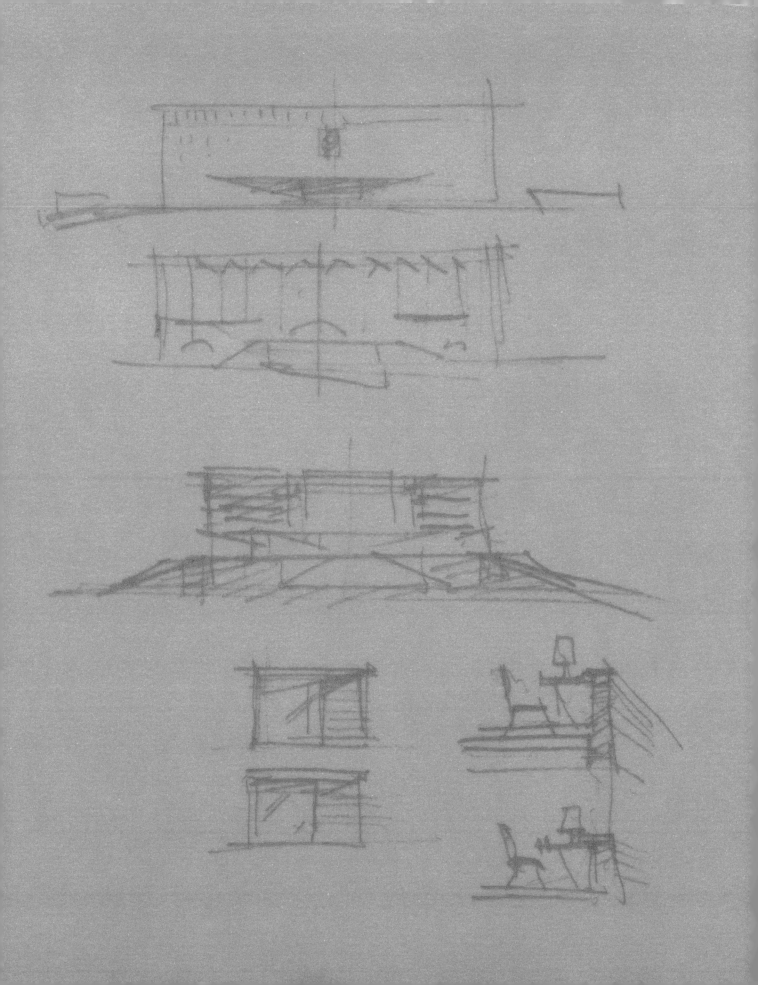

图书馆位于东北大学新校区两条主要空间轴线的中心交点，是新校区的标志核心建筑。建筑外观形态方正完整，与周边大体量教学楼尺度协调，同时利用"堆坡""墙"等手法延伸图书馆视觉高度。在面向两条校园轴线方向采用不同的形态处理策略，与周边建筑共同构成井然有序的校园空间。

新图书馆注重对东北大学悠久历史人文精神的传承，设计以杨廷宝先生设计的东北大学原北陵校区图书馆为文脉之源，外墙选用清水砖砌筑，从立面比例尺度、材料搭配到窗口门洞构造处理，都回应这一历史建筑的精彩细节，呈现东北大学厚重内敛的校园气质。建构表达、材料选择以符合当地气候、顺应材料逻辑为原则；大开大合、虚实对比的组合形成图书馆规整而有活力的外观形象。

平面布局重视开放高效、空间共享，公共阅览部分由南、北两侧进入，直达二层中央大厅，各类阅览空间围绕大厅层层展开，多样性的阅读空间可满足各类学习需求。

As the landmark of the new campus, the library is located at the central intersection of the two main spatial axes of the new campus of Northeastern University. The appearance of the building is square and complete, coordinated with the scale of the surrounding large teaching buildings, and increase the visual height of the library. The façades facing different axis are designed with different vocabularies to emphasize the space sequences of the campus.
Inheriting the long history and humanistic spirit of Northeastern University, the library continues the context of the original Beiling Campus Library of Northeastern University designed by Mr. Yang Tingbao. The design of the new library's brick walls echoes the wonderful details of the classical building in proportion, material collocation and tectonic treatments, presenting the dignified cultural atmosphere of the campus.
Openness and sharing are focuses of the functional layout. Students enter the public reading room from the north and south sides and reach the central hall on the second floor directly by large stairs. The reading spaces on each floor are attached around the hall to meet various learning needs.

东北大学浑南校区图书馆 · 2012—2018 · HUNNAN CAMPUS LIBRARY OF NORTHEASTERN UNIVERSITY

传承：向校园历史的致敬，不仅是材料、色彩的沿用，更是厚重内敛的气质、"粗粮细作"的讲究、务实创新的精神传承。

关爱：开敞的空间、适宜的尺度、舒服的家具、温暖的灯光、安静的氛围，为师生营造值得依恋的场所，无论寒冬还是盛夏。

精神：理性务实、稳重开放、简洁大气，是工科类大学求实精神的象征。

共享、交流：读书、讨论、休憩、集会、展览，各类空间都连接在一起；同学、老师、领导、访客各类人群都在一个屋顶下相遇。

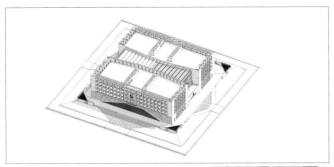
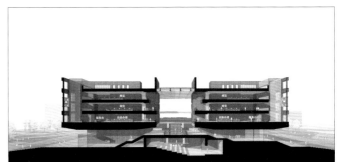

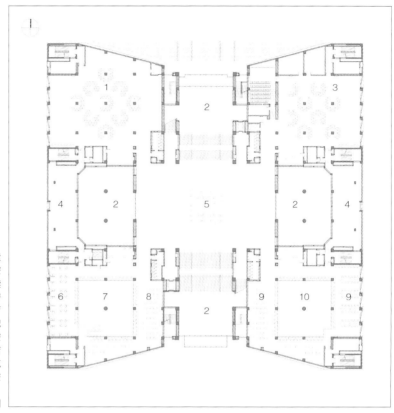

1. 网络文献检索
2. 门厅上空
3. 多媒体休闲阅读
4. 备品库房
5. 中央大厅
6. 咖啡书吧
7. 24小时阅览上空
8. 书店
9. 报刊新书阅览
10. 教师研修室上空

二层平面图

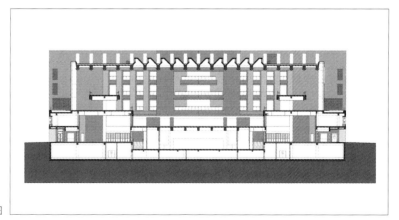

剖面图

LAND-BASED RATIONALISM III *041*

东北大学浑南校区图书馆 · 2012—2018 · HUNNAN CAMPUS LIBRARY OF NORTHEASTERN UNIVERSITY

东北大学浑南校区图书馆 · 2012—2018 · HUNNAN CAMPUS LIBRARY OF NORTHEASTERN UNIVERSITY

东北大学浑南校区图书馆 · 2012—2018 · HUNNAN CAMPUS LIBRARY OF NORTHEASTERN UNIVERSITY

设计随笔

东北大学浑南校区图书馆项目在整个建筑设计阶段始终坚持"适宜"设计——这个适宜不仅体现在适应场地环境、东大历史、校园文化等方面的回应上；也体现在设计对业主诉求、施工环境、地域习惯等非传统设计领域，为全局考虑的回应上。

砖作为建筑立面的主材，是在校园规划中已经确定的基调——深沉厚重的红砖既契合东北房子的厚重质朴，也能表达校园建筑的严谨沉静，更是新旧校区之间可以传承的媒介。我们要在这个四方的红砖体量上做减法，这一过程就像拿着一方石印做雕刻。

一座好的砖建筑，必须从设计一块砖开始，这也是保证砖建筑完成质量的决定性环节。在方案设计基本定型后，我们由平面上8400mm的方格轴网定义出清水砖墙在正交方向需形成262.5 mm的基本模数，进而推定得到本项目的清水砖尺寸——整砖254 mm × 123 mm × 58 mm、半砖123 mm × 123 mm × 58mm。整个图书馆的内外墙体都是用这一种尺寸的砖砌筑而成，几乎不产生碎砖切砖，统一后用量有保证，厂家开模烧制不成问题，清水砌筑效果因此得到保障。

平面斜角的设计也在清水砌筑这最细节的层次上得到了体现，并增加了清水砖墙的细节和特色效果。斜角处的对角错开砌筑方式使整面砖墙多处产生了细小有序的阴影细节，在最小的层次上，这种角部处理将建筑上"大"的尺度和线条化整为零，再次呼应了整体形态的雕塑感。注2

乡情的抚慰
The soothing touch of nostalgia

昆山西部医疗中心一期 · KUNSHAN WESTERN MEDICAL CENTER , PHASE 1
设计 Design 2014 · 竣工 Completion 2022

地点：江苏昆山 · 用地面积：125 976平方米 · 建筑面积：202 434平方米
Location : Kunshan, Jiangsu · Site Area :125,976m² · Floor Area : 202,434m²

合作建筑师：陈一峰、赵强、崔磊、段建军、袁媛、谭新颖、郭祺伟、何继舜、
　　　　　　王玉嘉、蔡建滨、林立川、梁桂荣、梁凯雁、叶松源
Cooperative Architects : CHEN Yifeng, ZHAO Qiang, CUI Lei, DUAN Jianjun, YUAN Yuan, TAN Xinying, GUO Qiwei, HE Jishun,
　　　　　　　　　　　　WANG Yujia, CAI Jianbin, LIN Lichuan, LIANG Guirong, LIANG Kaiyan, YE Songyuan

合作机构：新加坡CPG咨询私人有限公司
Cooperative Organization: CPG CONSULTANTS PTE LTD

策　略：格局、借形、色彩、生活、意境

摄影：徐晓飞、崔磊、刘晶
Photographer: XU Xiaofei, CUI Lei, LIU Jing

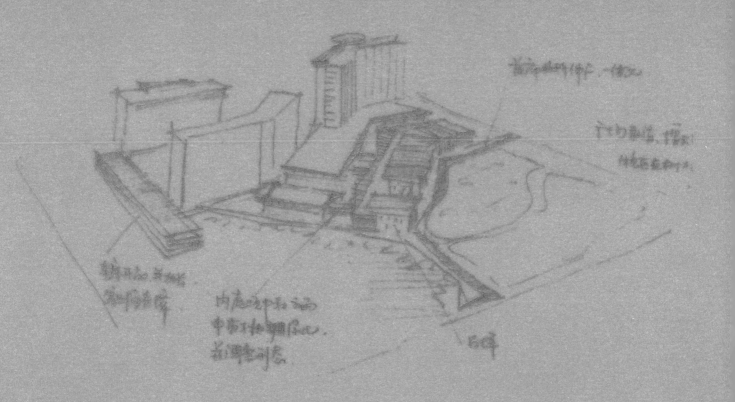

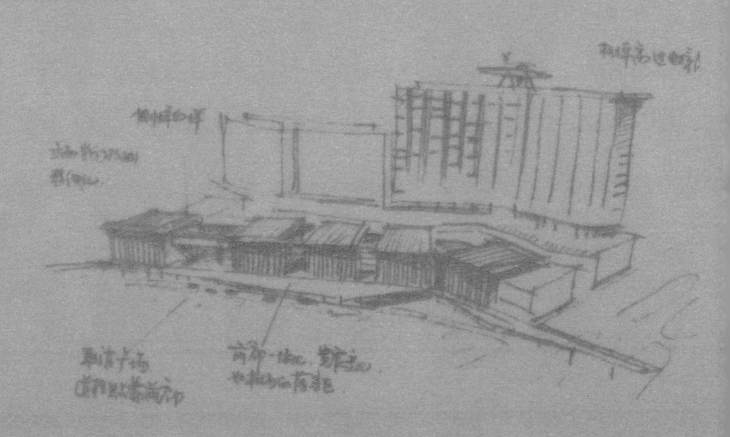

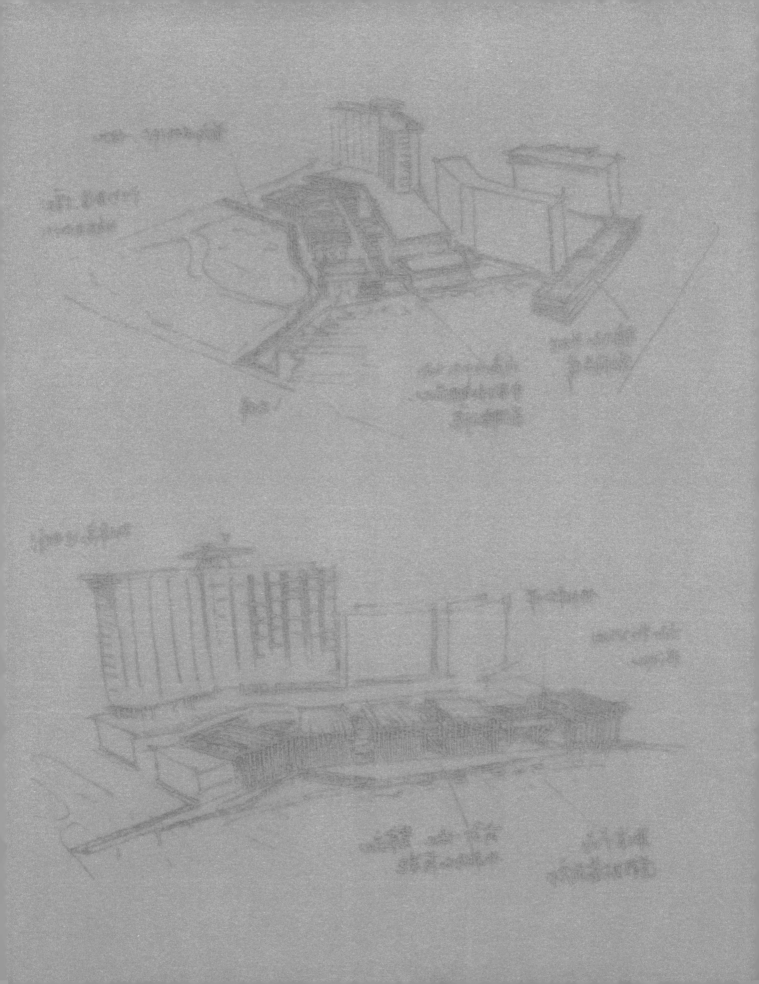

昆山西部医疗中心地处昆山巴城镇，是一座1200床规模的综合中医医院，整体布局采用45°转角设计，获得充足日照的同时形成迎候患者的城市花园。通过植入风雨连廊和地下通道，打造舒适步行系统，连接公交车、地铁接驳站和院区主入口，使院区无缝接轨城市交通网络，拉近医院与城市、与人的距离。设计将建筑体量打散，化大为小，引入传统水乡街巷的格局，以室外医疗街连接门诊和医技两部分，提供拥有自然通风采光的室内环境和中心景观的休憩空间；门诊区拆成若干模块，穿插室外平台、骑楼片墙与连续性的景观庭园，消解建筑压迫感，营造小尺度宜人空间，传统园林文化在这里演化为服务于医患的人文设施。风雨连廊与门诊飘檐作为车行入院的落客区，也为患者提供遮风避雨的步行和候诊空间，交通的可达便利及患者体验感被高度重视。住院部设有架空层及屋顶花园，为医生、住院患者及家属提供了宜人室外空间。建筑色彩、室内设计中提取传统元素，营造安静祥和、具有地域性和亲切感的空间氛围。病房采用"一床一窗"模式，使三人间病房获得单人间品质。这座位于江南园林中的医院，将成为承载地域文脉的城市名片，取于城市，馈于城市，为人们带来健康、自然、舒适的就诊体验，使医疗环境成为社会活动的一部分。

Located in Bacheng Town of Kunshan, the medical center is a comprehensive TCM hospital with a capacity of 1,200 beds. The overall layout featuring 45-degree angles is adopted to obtain daylight and form an urban garden with a welcome gesture for the patients. A pleasant pedestrian system was established with roofed corridors and underground passages, connecting transport stations and the main entrance of the hospital. The layout was inspired by the layout of traditional Chinese waterside towns to form a modern large-scale hospital with scattered functional blocks. An outdoor medical street connects the outpatient sector and medical technology sector, providing a space with natural ventilation, daylighting and central landscape. Traditional elements are embedded in architectural colors and interior design to create a quiet, peaceful, regional and intimate space for patients. Accessibility and patients' experience are highly valued in all areas of the hospital. The layout of inpatient wards features the model of "one bed with one window", offering experiences similar to that of single-patient wards. The garden-like hospital will serve as a landmark of the city, giving back to the city by providing healthy and natural experiences, making the medical environment an integral part of the city's social activities.

昆山西部医疗中心一期 · 2014—2022 · KUNSHAN WESTERN MEDICAL CENTER, PHASE 1

格局：现代医院中引入传统街巷的格局，让患者体会到乡情的温度，紧张会变为松弛，陌生会变为亲切。

借形：借传统之形，化解建筑的体量，营造温馨的场所，表现中医的文化。

色彩：具有地域性的色彩不仅是为了表达建筑的地域性，更希望让患者熟悉而亲切，找到家的感觉。

生活：沉痛也是生活的一种状态，医院也是生活的一个场所。为患者的设计，应关注到从生理到心理的需求，这是设计医院的基本点。

意境：意境是人在身心体会中的感知，中医是患者用心体会医术和医药的调理方式，为中医院设计要注意创造有助于唤起患者感知的氛围。

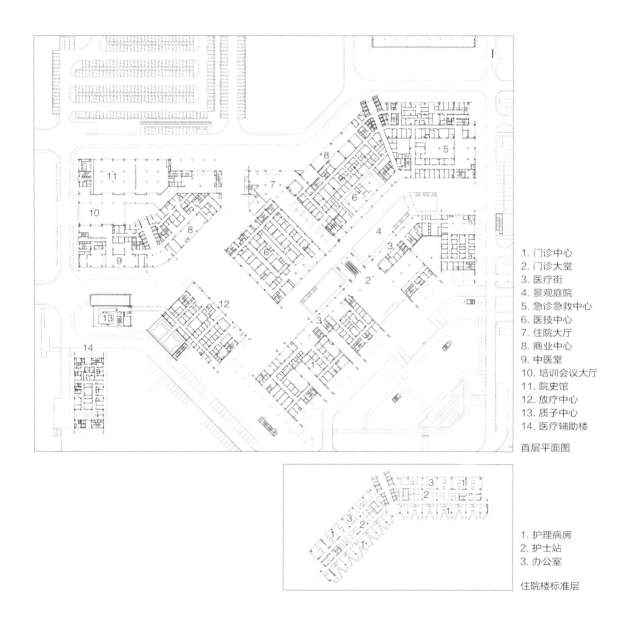

1. 门诊中心
2. 门诊大堂
3. 医疗街
4. 景观庭院
5. 急诊急救中心
6. 医技中心
7. 住院大厅
8. 商业中心
9. 中医堂
10. 培训会议大厅
11. 院史馆
12. 放疗中心
13. 质子中心
14. 医疗辅助楼

首层平面图

1. 护理病房
2. 护士站
3. 办公室

住院楼标准层

昆山西部医疗中心一期 · 2014—2022 · KUNSHAN WESTERN MEDICAL CENTER, PHASE 1

昆山西部医疗中心一期 · 2014—2022 · KUNSHAN WESTERN MEDICAL CENTER , PHASE 1

昆山西部医疗中心一期 · 2014—2022 · KUNSHAN WESTERN MEDICAL CENTER, PHASE 1

昆山西部医疗中心一期·2014—2022·KUNSHAN WESTERN MEDICAL CENTER, PHASE 1

设计随笔

自2020年暴发的新冠疫情对经济和社会产生了深远影响，也暴露了我国医院建设量不足的短板，医疗行业面临大规模的建设需求。同时，疫情改变了人们的健康意识，引发了对未来生活模式的深刻思考。人们健康意识的逐渐增强对医疗建设也提出了新的要求，医院不仅要满足求医问药的基本功能，更是与日常生活息息相关的重要场所。

设计目标

项目位于昆山老城区西部，区域总体定位为"集居住生活、医疗卫生、商务办公及文化休闲于一体的城市综合功能区"。基地东侧是新江南商业用地及生态公园，西、北侧是多、高层住宅区，南侧隔主干道相望昆山市公共卫生中心，周边生态环境较好，居住氛围浓厚；紧邻地铁S1线与公交车站，交通便利。

国内正处于医疗项目大规模建设中，但多数医院设计偏重满足功能需求而缺失本土文化，对人和环境的关注不够，对比国际医疗建筑水准尚有一定差距。昆山西部医疗中心（全期）作为2000床位、总投资30亿的大型医疗项目，如何体现人文性与本土性？人们置身于其中会获得怎样的感受？对人的日常行为与生活模式会产生哪些影响？为人与社会营造怎样的健康医养环境？

我们的答案是，要设计一座处于江南园林中、承载地域文脉的医疗建筑，它将作为城市名片取于城市、馈于城市，为人们带来健康自然、舒适的就诊体验，并使医疗环境成为社区活动空间的一部分。

园林化

项目整体布局采用45°转角设计，既避开京沪高铁的噪声，也获得了充足的日照，并释放出大量的景观绿地，形成了迎候患者的城市花园，与西南角的康养花园、西侧蜿蜒的滨水栈道及北侧的医患花园串为一体，让医院处于郁郁葱葱的花园环境之中。

从空间上，建筑前檐设计了风雨连廊，既可作为平时车行入院的落客区，也提供了遮风避雨的步行交通空间，还可以作为患者的候诊空间，用于医院组织宣讲、医普等活动——传统园林文化在这里演化为服务医患的人文设施。

园林化布局没有止于门前，还渗透到内部空间。建筑本身是一座庞大的综合医院，在设计中将体量打散，化大为小，层次丰富：将门诊区拆成了若干个模块，并穿插了进退有致的室外平台、鳞次栉比的骑楼片墙与连续性的景观庭园，消解了建筑的压迫感，尊崇了水乡的图底关系，营造了小尺度的宜人空间。

医院内街的下沉景观庭院是门诊和医技两区共享的空间，与内街形成视线呼应和丰富的垂直空间景观，植被掩映下的景墙构建出既独立又连续的园林空间形态。

住院部主楼设有架空层及屋顶花园，为医务工作者、住院患者及家属提供了宜人的室外空间。架空层通风避雨，可容纳休憩就餐、运动健身、儿童活动、独处休憩等多样行为。在屋顶花园上可以近距离看到高层住院楼的主立面，通过坡道建立花园与架空层之间的视觉沟通，将折廊堂榭的园林意向以立体游廊的形式传递。

人性化

在设计中传递"患者即宾客"的理念，为就医人群提供如酒店般的高品质服务体验。

交通的可达便利被高度重视，人们或从地铁连接通道步入商业环境，乘扶梯方便地上行至主入口，或乘车到达后在深檐雨棚下落客。风雨连廊将公交车站、地铁出入口与门诊广场无缝连接，并贯穿整个院区，提供交通引导性的同时，也为人们提供可遮阳避雨的等候、休憩、漫步空间。

门诊大堂采用逐层升高的木格栅吊顶，提取传统披檐元素，丰富空间层次的同时获得更好的采光与通风。从减轻病患心理压力角度出发，室内设计减少不必要的装修造型，以粉墙灰套、竹色帘幕营造安静祥和的空间氛围，并将具有中国文化象征的书法元素融入标识设计。

拥有自然通风采光的医疗街贯通各个候诊休憩空间，改变了传统的封闭在室内座椅间的候诊体验，结合江南宜人气候，与中庭室外游廊相连通，既能缓解患者的紧张情绪，也使候诊行为更为舒适。

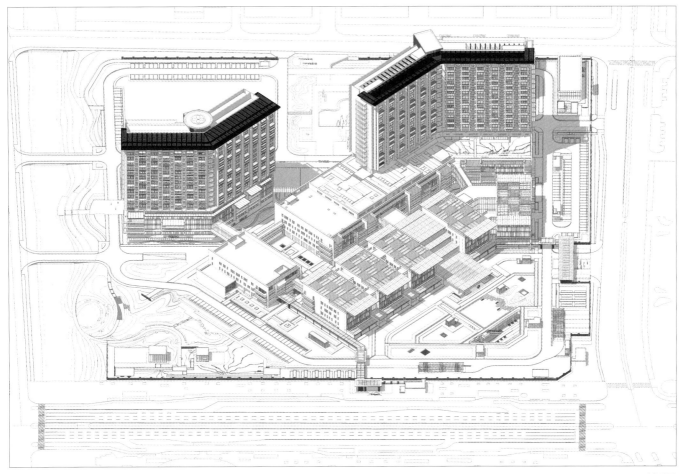

轴测图

在住院门厅设计了宾至如归的居家氛围,让住院患者产生亲切感。病房设计利用10.5m的结构柱跨,采用"一床一窗"模式,每一张床位都拥有独立的窗,使三人间病房获得了单人间品质。病房阳台采用全玻璃高栏板,视线可无遮挡地望向院区景观,下方设有透气格栅使空气流通。病房层两端设置病人活动室,在提供晾晒空间的同时,可以让患者在住院休养期间拥有全视野的景观资源。

在急诊区,除了宽敞的通道和充足的急救设施外,还为陪护者提供了园林化的等候区,温馨的景观可以缓解人们的情绪和压力。注3

戈壁中的聚落
A settlement in the Gobi

敦煌市公共文化综合服务中心 · DUNHUANG PUBLIC CULTURE COMPREHENSIVE SERVICE CENTER
设计 Design 2013 · 竣工 Completion 2017

地点：甘肃敦煌 · 用地面积：20 000平方米 · 建筑面积：19 936平方米
Location : Dunhuang, Gansu · Site Area :20,000m² · Floor Area : 19,936m²

合作建筑师：吴斌、辛钰、崔剑、郑虎、董静文
Cooperative Architects : WU Bin, XIN Yu, CUI Jian, ZHENG Hu, DONG Jingwen

策　略：借形、格局、肌理、色彩、符号

摄影：张广源
Photographer: ZHANG Guangyuan

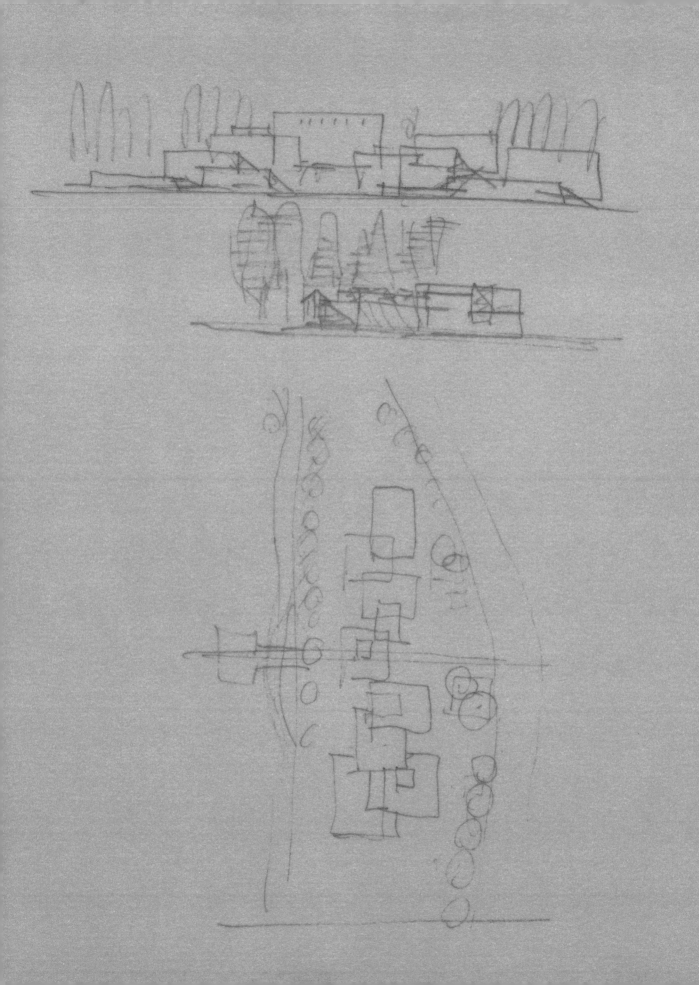

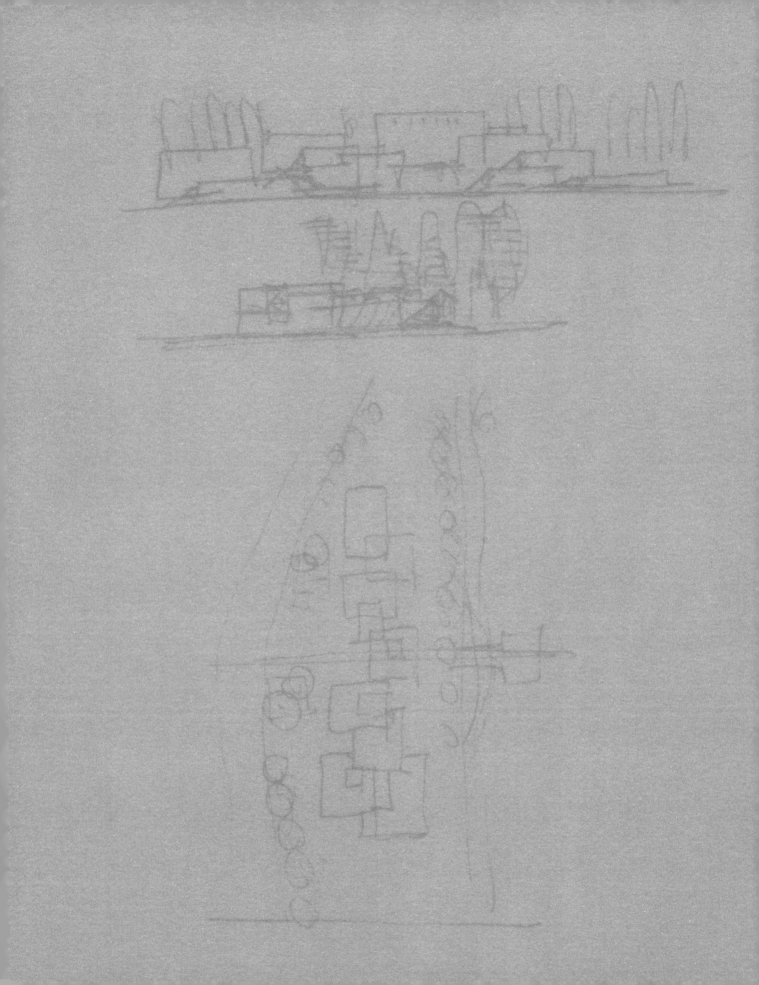

　　本项目以"聚落"为设计理念,在统一模数的控制下,将博物馆、图书馆、文化馆、档案馆等工程整合起来。以当地民居常见的院落形态进行布局,提供自然采光和通风,有利于节能,同时增加空间的层次。建筑体块高低错落,既减小建筑的尺度,同时形成不同标高的活动平台,为使用者提供开放的交流和沟通的场所。外立面根据内部功能需要尽量减小开窗,小方窗的组合方式进一步强调立面的秩序感,如同当地葡萄晾房镂空的方洞,产生较深的阴影。立面材料用土黄色的砂岩,土黄色来自鸣沙山、戈壁的颜色,表面的肌理在光线下产生犹如砂砾般粗犷的质感。这组台地式的土黄色砂岩建筑,在光影的刻画下,以高低错落的"聚落"形态,形成一个尺度亲切、层次丰富、开放立体的公共空间。

With the concept of clusters and the framework of modules, various facilities, including a museum, a library, a cultural center and an archive are integrated into one building. A layout dominated by courtyards can facilitate natural lighting and ventilation while adding to the diversity of the space. The different heighted volumes have endowed the building with a proper scale and platforms on different altitudes for communication. The neatly arranged small windows, like the holes on the walls of local grape-drying houses, cast long shadows. The sandstone with earthy color on the façade gives the building a rough texture. The stepped volumes form an upward route to rooftop platforms at various altitudes. The outdoor gallery dominates the highest point, where visitors can have a panorama of both the city and the undulated Mingsha Hill.

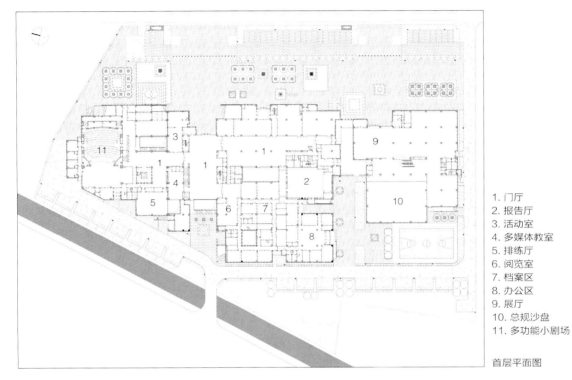

1. 门厅
2. 报告厅
3. 活动室
4. 多媒体教室
5. 排练厅
6. 阅览室
7. 档案区
8. 办公区
9. 展厅
10. 总规沙盘
11. 多功能小剧场

首层平面图

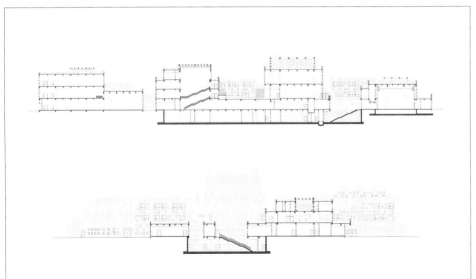

剖面图

敦煌市公共文化综合服务中心 · 2013—2017 · DUNHUANG PUBLIC CULTURE COMPREHENSIVE SERVICE CENTER

设计随笔

本项目考虑的核心问题是如何以一个开放的姿态，将几个馆的功能有效地组织并整合起来，既能满足当地老百姓的日常使用，又能结合每年的文博会举办各种文化活动，成为敦煌对外文化交流的重要场所。而这样的一组建筑还应恰当表达敦煌的自然和地域文化特色——文化性的表达是这个项目解题的关键。

从卫星图上看，敦煌实际上是茫茫戈壁的一个小绿洲。整个城市由发源于祁连山的一条河流冲积形成，城区基本上沿着东西向和南北向的两条主要轴线建设发展。文化中心的基地位于市博物馆南侧约500m，再往南几公里即到鸣沙山月牙泉。基地呈不规则四边形，北窄南宽，东邻敦月公路，交通便利；南北为农家林地，挺立着一排排非常精神的穿天杨，还有大量的果树；西侧有一条灌溉渠，是敦煌农业的生命线，渠边绿树成荫；再往西就是成片的绿油油的葡萄园。如果在基地内登高，南可欣赏鸣沙山，北可俯瞰市区。

在方案构思之初，我们曾提出这样一个思路：打破几个馆的界限，将所有功能糅到一个大的建筑里，公共区域完全共享，各个功能区互相联通，整个建筑如同一艘巨舰一样，以超尺度表达一个文化综合体的形象。但是在初步讨论时，大家在赞赏之余表达了这样一个疑虑，为老百姓服务的这样一组文化建筑，应当是宏伟的还是亲切的？宏伟的建筑纪念性很强，但容易产生距离感，而亲切的建筑容易让老百姓接近，更接地气。于是，我们改变了策略：以小尺度的体块，在统一模数的控制下将不同功能整合到一起，形成一组掩映于绿树之中的聚落形态，而小尺度更适合敦煌这样一个小镇的尺度。

我们在东侧留出城市广场，为大型文化活动留出场地，并结合几组小型绿化供人们休息。在基地中心的位置原本设想干脆直接架空，能一眼望穿看到水渠边上的大柳树，提供一个开放的、东西贯通的城市空间，但是考虑冬季很冷，在大家的建议下还是加上玻璃变成了室内公共空间，作为整个建筑的"客厅"。在大厅里，提供接待、咨询等综合服务功能，往上是美术馆及公共展厅，往下是非遗中心，往北通往文化馆和小剧场，往南通往图书馆和档案馆。各馆门厅旁边设置院落，让自然光洒进室内。各馆主体呈方形体量，中间再掏出内部中庭，于是形成主厅—门厅—院子—中庭—功能房间的空间序列，这是水平方向的一条空间线索。

另外一条空间线索则往上引导，通过台阶到达不同高度的屋顶平台。在东侧主入口两侧设置大型台阶将人们往南或往北引导到二层开放平台，北侧平台作为文化馆的入口和小剧场二层的疏散口，南侧平台作为图书馆、档案馆的入口，同时也连接办公区域和城市规划展览馆。中间美术馆及公共展厅部分甚至可以再继续围着主体拾级而上，到达屋顶平台的空中室外展场，北眺城市小镇风光，南望鸣沙山之曼妙曲线。图书馆区域也可以让读者在阅读之余登顶远眺，吸收大自然的营养。

院落是这两条空间线索的交汇点，就像当地民居院落一样，既可以躲避风沙，又能获取自然采光和通风。敦煌属于沙漠气候地区，空气干燥，虽然太阳辐射强，但是只要有遮荫通风的地方就会很凉快，所以我们通过院落、中庭、屋顶构架等方式形成采光通风和遮阳，尽可能不用人工空调，达到绿色节能效果，也能大大降低运营成本。

在高高低低的体块、错综复杂的平台和大大小小的院落组合中，建筑的功能、流线、形态和空间达成统一。丰富的形体让外立面的处理变得相对比较放松。根据内部功能，需要开窗的开窗，不需要开窗的就是实墙。图书馆开大窗，档案馆开小窗，剧场就是一个封闭的大盒子，在太阳辐射很强的沙漠气候中有利于节能。大小方窗的组合方式化解了单一体块的生硬感，同时也强调出立面中某种模数关系，如同当地葡萄晾房砖块之间镂空的方洞，产生的阴影让建筑变得更加立体。立面材料用的是从云南运过来但非常便宜的土黄色砂岩，在周围绿树的衬托下，表达出戈壁和绿洲的自然色彩搭配。砂岩表面有天然肌理和色差，甲方最初担心会不会太"花"，在我们的坚持下挂上去，发现这些色差就像点彩画一样，在大实大虚的墙面上反而形成丰富的表情，再加上毛面和纹理，在光线下产生犹如砂砾般的质感，生动而自然。

　　于是，这样一组台地式的土黄色砂岩建筑，在光影的刻画下以高低错落的"聚落"形态，形成一个尺度亲切、层次丰富、开放立体的公共文化活动空间，如同茫茫戈壁中矗立的烽燧，如同落日余晖中金色的玉门关，如同烈日下沧桑的洞窟崖壁，沐浴并雕刻着时光。

　　项目建成后的这两三年里，当地老百姓非常喜欢到这里来。年轻人与朋友周末相约图书馆、看书聊天；退休的老人们可以在专门的老年人阅览室看书，也可以在文化馆里琴棋书画、唱歌跳舞；孩子们也特别喜欢在儿童阅览区里看书、做手工，在舞蹈教室里跳芭蕾。在旅游旺季和文博会期间，学术报告厅里从全国各地请来的名家不断；小剧场里每周都有音乐会和各种演出，场场爆满；美术馆和非遗中心各种画展和活动层出不穷。这组建筑给整个城市注入了新的活力，成为敦煌一处新的"文化聚落"。注4

营造有风情的场所
A space with Tibet charm

京藏交流中心 · BEIJING-TIBET COMMUNICATION CENTER
设计 Design 2017 · 竣工 Completion 2020

地点：西藏拉萨 · 用地面积：9 048平方米 · 建筑面积：39 856平方米
Location : Lasa, Xizang · Site Area : 9,048m^2 · Floor Area : 39,856m^2

合作建筑师：康凯、吴健、单晓宇、李俐
Cooperative Architects : KANG Kai, WU Jian, SHAN Xiaoyu, LI Li

策 略：嵌入、隐喻、观景、标识

摄影：张广源、李季
Photographer: ZHANG Guangyuan, LI Ji

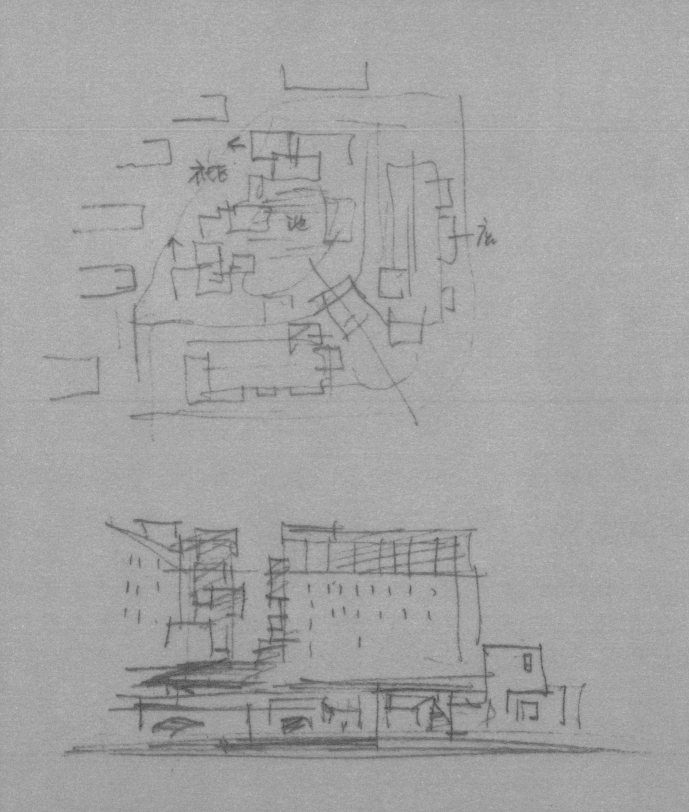

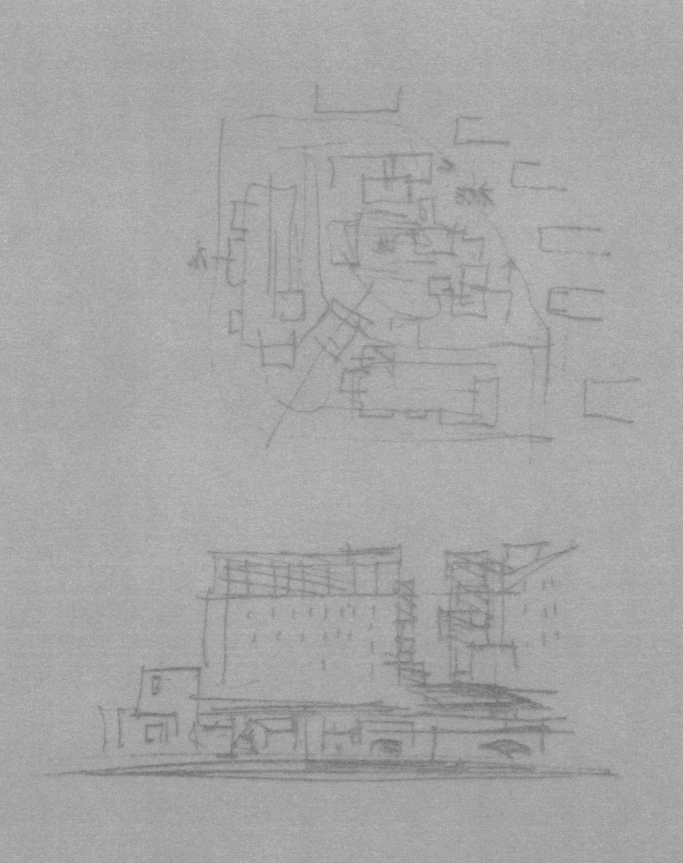

项目基地位于拉萨市东部新区，南侧紧邻拉萨河，向西可以遥望布达拉宫，作为北京与拉萨对口支援工作的根据地，为京藏两地经济文化交流提供服务平台。设计希望能以现代建筑语汇表述与藏区文脉传统的传承与共生，以谦逊友善的姿态融于城市肌理和自然之中。设计取自藏式宗堡建筑的空间构型，以大小不一的方楼体量为基本构成单元，结合功能进行灵活组合。

在公共区营造延伸递进的空间序列，景观庭院用类转经回廊串联起主要功能空间，回廊亦廊亦厅，营造丰富有序的多重空间体验。外墙主要采用劈开石湿挂，将藏族民居砌筑墙体常用的石头与当代石材加工工艺、构造做法相结合，形成与远山河滩遥相呼应的粗犷肌理。通过对藏式建筑中檐口、门、窗等元素符号的系统化提炼，以现代材料构造加以抽象演绎。

Located in the Eastern New Area of Lhasa, the site of the center borders Lhasa River on its south and faces the Potala Palace to its west across a distance. As the base of Beijing-Lhasa Aid Project and a platform for Beijing-Tibet communication, the center stands in harmony with the urban context and nature with a humble posture, expressing its inheritance and co-existence with Tibet culture.

Derived from the prototype of Tibetan castles and composed of various sized square volumes as its fundamental units, the building has presented a spatial sequence at its public sector, with corridors in the landscape courtyard connecting main places. Wet-hung split stones dominate the façade, where a unique style is presented with modern building technology applied to traditional materials.

京藏交流中心 · 2017—2020 · BEIJING-TIBET COMMUNICATION CENTER

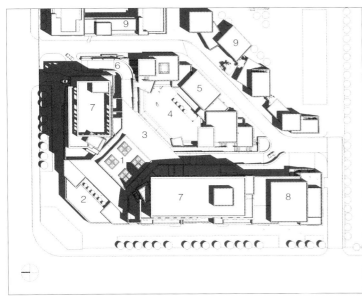

1. 大堂
2. 援藏成果展厅
3. 落客区
4. 水院
5. 餐厅
6. 汽车坡道
7. 客房楼
8. 商业
9. 公寓底商

总平面图

剖透视图

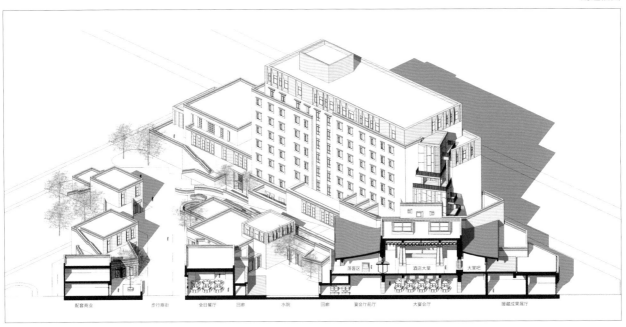

京藏交流中心 · 2017—2020 · BEIJING-TIBET COMMUNICATION CENTER

京藏交流中心 · 2017—2020 · BEIJING-TIBET COMMUNICATION CENTER

筑就精神的高地
Building the spiritual highland

铁道游击队纪念馆 · THE RAILWAY BRIGADES MEMORIAL
设计 Design 2018 · 竣工 Completion 2019

地点：山东枣庄 · 用地面积：26 398平方米 · 建筑面积：11 500平方米
Location : Zaozhuang, Shandong · Site Area :26,398m^2 · Floor Area : 11,500m^2

合作建筑师：邢野、高凡、金爽、周益琳
Cooperative Architects : XING Ye, GAO Fan, JIN Shuang, ZHOU Yilin

策 略： 借形、符号、精神、艺术

摄影：张广源
Photographer: ZHANG Guangyuan

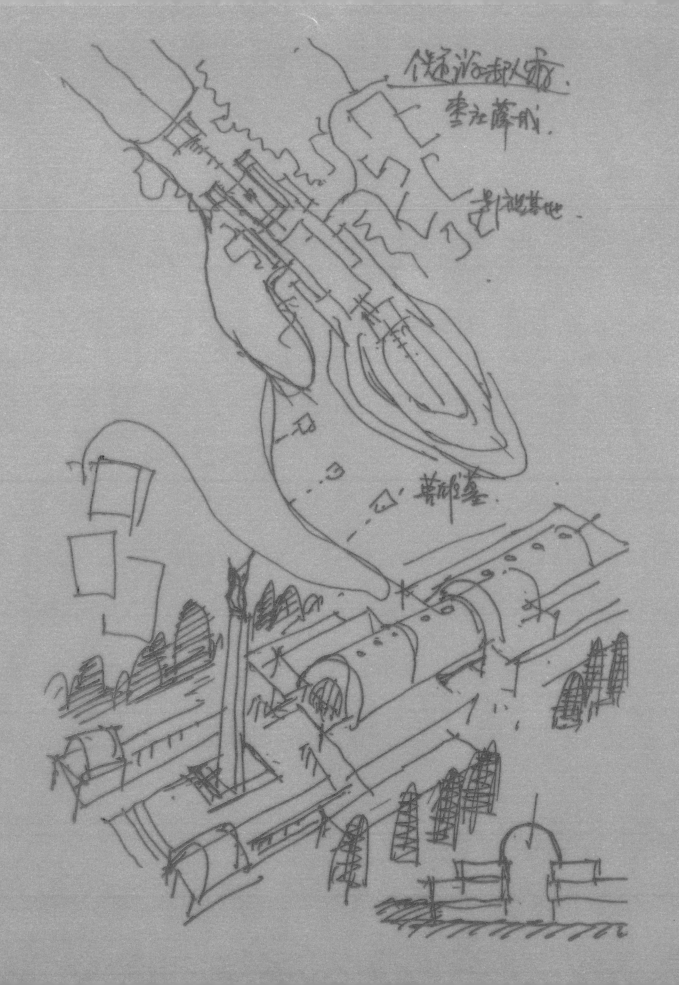

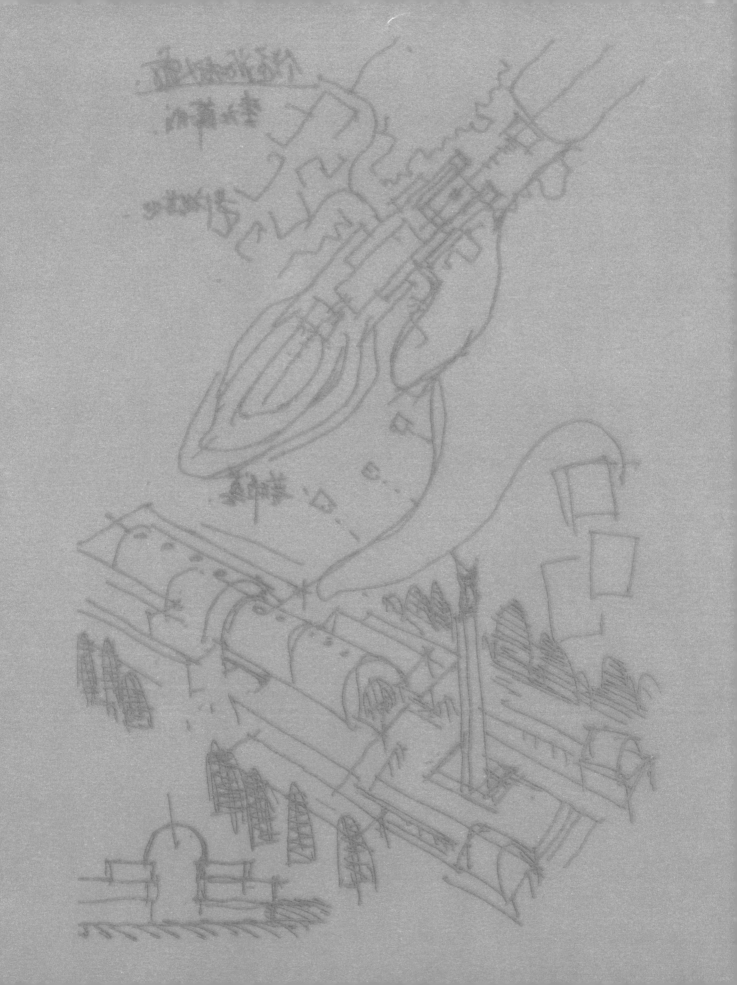

　　铁道游击队纪念馆位于枣庄临山纪念园内，在处理复杂场地问题的同时，也为建筑形体营造以及建筑内部和外部的流线处理带来了机会。建筑、碑廊与纪念碑广场平台共同形成西低东高的整体态势，西侧建筑尽量压低以突出纪念碑的主体性，两侧碑廊环抱纪念碑广场，东侧建筑体量呼应整体山势逐级上升，并通过屋顶绿化与山体融为一体。利用局部下降的场地条件架空设置，在地面上让开连接南北的通路，并将参观结束的人流继续引导至纪念园内其他节点。拱顶覆盖相互连通的各主题空间，真实的火车和铁轨片段被置于纪念馆主轴线上，并通过空间的互相借用形成丰富、立体、有纵深感的空间体验。

Located at Linshan Memorial Garden of Zaozhuang, the Memorial Hall of Railway Guerrillas features a design that has overcome difficulties posed by complex site conditions. The buildings, the corridors of steles and the monument square platform have formed an overall volume that descends from the east to the west, where the buildings at the west remain lower to highlight the significance of the monument. Corridors of steles embrace the monument square, and the buildings at the east part of the site ascend in accordance with the mountain's shape. Part of the buildings are built on stilts to provide north-south access across the site.

铁道游击队纪念馆 · 2018—2019 · THE RAILWAY BRIGADES MEMORIAL

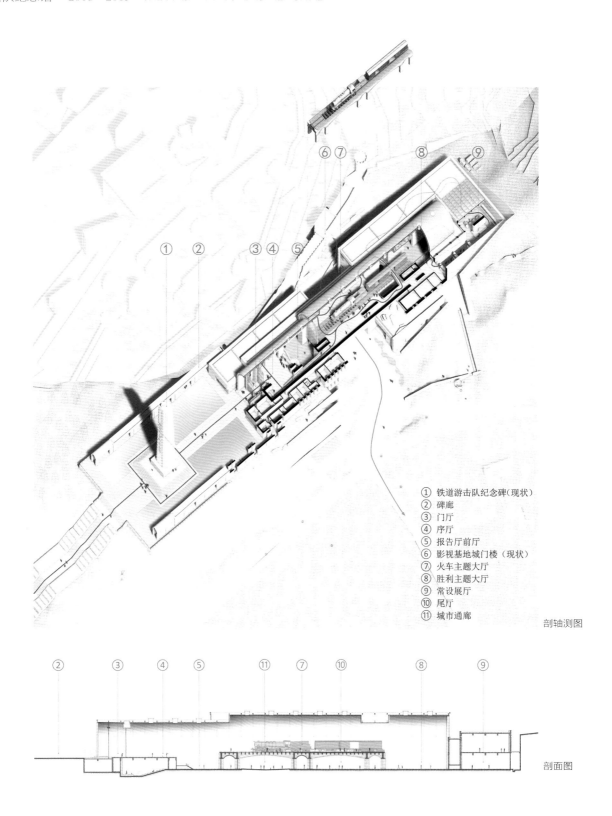

① 铁道游击队纪念碑（现状）
② 碑廊
③ 门厅
④ 序厅
⑤ 报告厅前厅
⑥ 影视基地城门楼（现状）
⑦ 火车主题大厅
⑧ 胜利主题大厅
⑨ 常设展厅
⑩ 尾厅
⑪ 城市通廊

剖轴测图

剖面图

LAND-BASED RATIONALISM III *081*

铁道游击队纪念馆 · 2018—2019 · THE RAILWAY BRIGADES MEMORIAL

LAND-BASED RATIONALISM III 083

设计随笔

铁道游击队纪念馆选址在枣庄临山纪念园内，首先的挑战就是如何在已经成形的纪念园内补充一座纪念馆。最终的选址位于山腰处纪念碑以东与影视基地南入口临近的停车场空地上——一处混乱而且没有被充分利用的空间——如此操作既可以最小化地侵占自然资源环境，又可以最大化地顺应原有纪念轴线：依托于现存的纪念碑，在空间序列上先碑后馆，合理的碑馆关系符合纪念礼仪，是对原有空间秩序顺势而为的结果。

纪念馆以水平性的体量匍匐于大地之上，西侧尽量压低，突出纪念碑的主体地位。在纪念广场的南北两侧，设置了廓清边界的碑廊，碑廊尺度如同横卧的纪念碑，与纪念馆一起形成了对纪念广场和纪念碑的环抱合围之势。广场经过改造，铺满黑色的砾石，强化了纪念的主题和场所的氛围。

纪念馆通过沿轴线的对称布局首先建立了强烈的纪念性，而后设计的复杂之处在于如何以对称的布局来面对复杂的场地条件。为此，我们将建筑形体策略性地分化为两部分，一是沿中轴线的金属筒拱，它的线性尺度和突出的形式特征进一步强化了纪念轴线；而另一个是轴线两侧错动上升的石墙方台，平面和剖面上的错动变化同时应和了外部场地和内部功能的复杂性。

从中心筒拱进入后，拱顶贯穿始终，覆盖着主要的主题空间。拱顶带来的中心性、层级展开的序列性、一点透视的纵深性都在宣示着绝对的纪念性，而在达到这种必要的纪念性之后，如何通过空间手段建构出一种"传奇性"，是铁道游击队纪念馆的内部空间营造的任务所在。

传奇性的表达主要隐含在对"铁道"和"游击"这两个内在主题的回应。首先是对"铁道"的回应。深色钢板拱顶覆盖下的幽暗冷硬的空间提示出一种类似隧道、矿洞一样的特殊环境，为进一步的空间叙事铺陈调性。在这样的空间背景下，引入火车和道桥作为重要的空间道具，将我们直接带入到铁道游击队员的战斗场景中。这列老火车在铁道游击纪念馆里不仅仅是一件展品，而且组织起整个纪念馆的空间，所有的空间叙事都围绕着它而展开。

不仅两侧展廊上有枣庄车站的场景图像，车内也用光电投影再现激烈的战斗场景，甚至桥梁轨道下采用玻璃托底，使人们在桥下的室外空间可看到庞大的车辆和铁轨，创造了独特的体验视角。

可以说，铁道游击队纪念馆一方面通过经典建筑语言建立起我们都比较熟悉的类似于社会主义现实主义的纪念性空间。然而另一方面，又希望通过回应铁道游击队纪念主题中所包含的特有的空间性和时间性，来突破对纪念性的一般性表达，借助内置空间要素的矛盾性将其打破，继而转化为一种具有更多空间向度和意义维度的超现实主义空间，以此接近建筑难以表达的、但历史让我们深切感受到的传奇性。注5

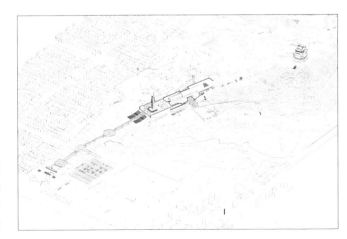

我最早知道枣庄这个名字还是上小学的时候，那时有本小人儿书叫《铁道游击队》，画得特别好，又是打鬼子的，很招小男孩儿的喜欢，都想方设法借来抢着看。后来大一点儿，又有了《铁道游击队》的电影，虽然是黑白的，拍摄质量和现在没法比，但那些演员精彩的演技和紧张的情节让我们百看不厌，也记住了刘洪和王强这两位传奇英雄的名字。真没想到时隔五十多年后，我能为这些英雄们设计一座纪念馆，对我来说这是一种荣幸和使命。

记得我第一次踏勘现场，登上位于城市中的英雄山，那不太高的山上遍植松柏，郁郁葱葱中矗立着十几年前建的纪念碑，碑后的廊子中有许多老将军们留下的题词。我们还瞻仰了位于侧面山腰处的英雄墓地，让我深切地感到枣庄人民没有忘记铁道游击队，我们的军队没有忘记这些老英雄。他们永远是祖国的骄傲，军人的榜样，代表着枣庄的红色文化精神！怀着这样的情感我们寻找设计的思路：建好的纪念碑不要动，但要重铺广场新建护廊，让它不要太单薄；纪念馆规模不算大，但要借山脊之势叠石筑墙，让它与山融为一体，如大地上隆起的脊梁；能收集到的展品不太多，但要以实体火车为主体创造主题空间场景和有体验性的展览效果；我还设计了山下的纪念园，用一座座石碑围成的广场如同大地上的花环，向枣庄的英雄们献上永恒的敬意。

这组纪念建筑景观建成以后，听说得到各级领导的肯定和百姓的喜爱，让我感到为枣庄做了一件特别有意义的好事。

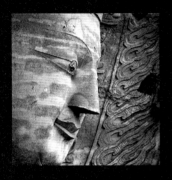

流去的辉煌
The flow of glory

大同市博物馆　**DATONG MUSEUM**
设计 Design 2010 ・ 竣工 Completion 2015

地点：山西大同 ・ 用地面积：51 556平方米 ・ 建筑面积：32 821平方米
Location : Datong, Shanxi ・ Site Area : 51,556m^2 ・ Floor Area : 32,821m^2

合作建筑师：时红、刘恒、邢野、吴健、冯君
Cooperative Architects : SHI Hong, LIU Heng, XING Ye, WU Jian, FENG Jun

策　略：隐喻、借形、色彩

摄影：张广源
Photographer: ZHANG Guangyuan

　　大同市博物馆选址于大同市新的行政文化中心御东新区，与大同古城隔河相望。博物馆位于新区核心位置，与东侧的音乐厅沿新区南北向轴线对称布置，北望行政中心，南侧为图书馆、美术馆，共同构成城市新的文化中心。博物馆的建设作为新区起步项目具有里程碑式的意义。设计从大同地区的龙图腾文化、火山群地貌特征中汲取灵感，两个弧形体量围绕中庭和庭院盘旋而起，并在统一回环结构内岔开不同的断面，为观展休息区引入光线和风景，展示空间随着非线性形态而变化，建筑端部的断面形式表达出力量感和对方硬汉字的形式隐喻。上下搭接的花岗石板覆盖着三维的建筑表面，并蔓延到形似圆壁的水池，有微妙色差的石板通过随机排布形成从下到上逐渐变淡的完整形式，加强了建筑锚固在场地并与天地浑然一体的气魄。

Datong Museum, located in a central position in Yudong New District, plays an important role in the construction of the area. The design has inherited the essence of the profound historical culture in Datong, while the architectural form draws inspiration from the time-honored dragon totem culture. Two simple arcs rising around the atrium have introduced light into the internal exhibition area. The exhibition space, which serves as the focus of the building, unfolds along the main parts of the building to make visitors feel as if they were going into a deep cave. The section of the building is exposed at the ends of the building. The non-linear envelope of the building is covered with granite slabs, which are also applied in a circular pool. The color of the slabs grow lighter from the bottom to the top, merging the building with the land where it belongs.

大同市博物馆 · 2010—2015 · DATONG MUSEUM

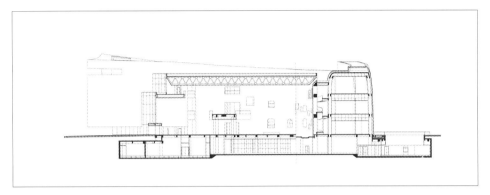

剖面图

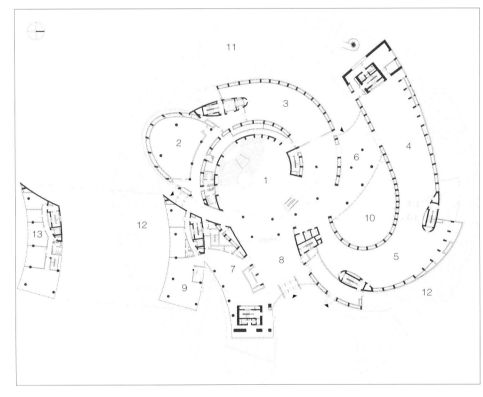

1. 大厅
2. 多功能厅
3. 远古恐龙化石展厅
4. 大同早期历史展厅
5. 唐代石刻展厅
6. 侧厅
7. 纪念品销售
8. 前厅
9. 咖啡茶饮
10. 室外庭院
11. 下沉庭院上空
12. 水池
13. 办公

首层平面图

大同市博物馆 · 2010—2015 · DATONG MUSEUM

设计随笔

说起大同市博物馆设计一定离不开耿彦波市长。记得那年他刚到大同任职不久，就邀请修龙院长（届时是中国建筑设计研究院的院长）和我去大同考察，他陪我们到老城里转，指着那些拥挤简陋的住宅说："那后面就是大同的老城墙，都淹没在这些破楼之中了，太可惜了。我一定要把城墙'亮'出来，修缮好，一定要重现大同的辉煌。这座古城中好东西太多了！"耿市长潜心研究过当年没能实现的"梁陈方案"，指着老城东侧的御河对岸说要在那边建座新大同，把老城的人口迁出来，才能拯救大同这座古城。市长的工作精神实在令人钦佩，下了决心说干就干。他请来各路专家团队领衔设计，从新城规划到老城修复，从安置住宅小区规划到每一个重要公共建筑设计方案，他都一一亲自把关。每天的工作汇报会排得满满的，从早开到深夜，一大早他还亲自下现场去检查督战，不满意就换人，就是大牌专家的方案不满意也要推翻重来，毫不含糊。我们先后设计的大同机场T2航站楼、大同市行政中心（太阳宫）、大同市博物馆等无不经历了多次汇报、不断优化才得以实施。

大同市博物馆设计之初我们去参观了老城里的老博物馆，陈旧的展厅中摆放着许多精致珍贵的文物，大多是北魏鲜卑人的遗物，记载了这个来自北方彪悍和传奇民族的历史。第一轮方案我们以大同的方城为模板，设计了比较周正平稳的方案。耿市长不太认可，觉得有些保守，建议我们去看看大同的九龙壁，又重点介绍了鲜卑这个游牧民族兴衰的历史，让我有所感悟。方城为静、游龙为动，流动的空间更适合北方民族征战出行的场景。而室外广场空间虽有轴线，并与其他三个公建成组布局，但每个建筑周边空旷，如雕塑般全视角呈现，没有正侧之分，圆形的体态、灵动的组合是更好的选择。另外建筑立面应该完整，呈现出具有标志性的超大尺度，所以将屋顶与立面一体化的筒形造型悬于水面，连办公区也用玻璃幕墙隐去层窗。平板石材深浅排版组合和鱼鳞式搭叠的工艺做法解决了曲面墙体的整体性问题，成本也比较低，防水也不会污染石材的密封胶，保证了它长久的品质。室内空间灵动有秩，展厅首尾相接，易于组织流线，厅厅之间可走到空中桥廊，既可俯瞰大厅也可外望远景，每个展厅的端部还有休息空间，让观众静思远望。

前不久我有机会重返大同，多年过去，博物馆早已成为大同人最喜爱的文化建筑，讲解员热情地介绍多年来市民游客对博物馆的赞扬和骄傲，让我由衷地感到欣慰。

关于象征　ABOUT SYMBOL

有人说国人喜欢象征，最明显的就是文字都是从象形文字演化而来。小时候，识字课老师就用各种图片讲字的构成，好听，好记。到如今还有书法家画字，有模有样，因字化形，因形生义，表达字之外的情趣。而"因形生义"更多见于日常，望山野岳峰可想到龟蛇龙虎，望院中小池可想到五湖四海。近些年建筑的形象更成了大众联想的焦点，各种"外号"层出不穷，"水煮蛋""大裤衩""小蛮腰""大钉子"。不管你领意不领意，"外号"一传出来便广泛回应，好名声或坏名声随之远扬，让领导都颇为紧张，早早提醒建筑师你这个设计像个什么？能不能先想个名字，正面引导？当然也有不少甲方为了吸引眼球，一定让建筑有个夸张的形态，像个什么形，甘冒被嘲笑、被贬讽的风险。看到这种现象有时真无语，是甲方太任性，还是大众趣味太低，还是建筑师没守住底线？

其实象征是建筑艺术表达的一种方式，历史上也不乏成功的象征主义的作品。比较广为人知的就是悉尼歌剧院，其象征贝壳或船帆都很鲜明地表现了悉尼港的海洋主题。尽管因为结构和造价的原因曾被诟病，但毫无疑问它作为澳大利亚的标志广受赞誉，绝对是物超所值的。我也曾进到里面参观，清水混凝土的结构和木制构件的精心组合，呈现出音乐的韵律，十分震撼，绝对是世界建筑史的大作！而民国时期吕彦直先生创作的南京中山陵也是一座象征主义的杰作，以古钟图形规划纪念陵园，象征中山先生为中华民族的觉醒敲响了警钟，不仅立意深远，也很好地处理了陵园和山地的关系，让无数后人沿大台阶上行，都有心灵的感悟。

从许多优秀的象征主义作品中，我感到要想用好这个手法，恐怕要把握好几点：1）立意高远、深邃，唤起人的心灵感悟和情感共鸣；2）原型要美，仿形适度，用建筑本体语言表现；3）象征的造型能引发内部空间的创新和体验，不是徒有其表。如果做到这三点，或者虽然达不到这个高度但也努力在这三方面有所追求，我想就不会流于庸俗、浅薄和造作；即便被人误读，也不至于遭人唾弃吧。当然，说回来，用象征的手法做设计的确比较难，容易出错、出丑，所以不是特别合适的条件下还是少用为好。对大多建筑而言，恰当地用建筑本体的建构语言还是正道。

原真：强调遗产的原状原地保护，不遮挡、伪装、再造。

修缮：对历史遗产的严谨、科学的保护性整修，不求完整、不能变新、不能做假。

保护：设计中碰到发掘出来的遗址遗产应尽量就地保护和展示，在可能的条件下纳入新建筑的空间中进行妥善保护和展示。

利用：对建筑遗产适度使用，前提是不因为使用功能去改变遗产的原真性，而让原真性成为新空间的有机组分。

展示：通过保留历史遗存和风貌，对相关历史信息做出介绍，使历史遗存发挥讲述历史、传播文化的作用。

对比：为了让后人能分清新建筑与历史遗迹的时空关系，设计中有意识地在新旧界面交接处采用不同的材质、色彩等处理。

协调：用高度、尺度、色彩、形式相似等策略建立新旧建筑的和谐关系，形成整体的协调。

对话：是一种新旧建筑或环境的呼应状态，比风貌协调更有开放性，可以在不同的时间、空间、主体之间展开。

消隐：利用嵌入地下或景观遮挡的办法让建筑消隐，以便保护遗产及其周边环境和场地的完整性。

风貌：与遗产环境总体上的协调控制策略。

控高：尽量压低建筑高度以减少对遗产环境的改变和干扰。原则上既要服从古建保护规划的要求，也应结合实际环境做视线分析，单一的指标控制并不符合实际情况，也难实施。

遗产保护

HERITAGE CONSERVATION

本土文化中最原真的就是历史遗产。历史遗产包括各类历史遗址、墓葬、历史建筑及街区等，也可以扩展到非物质文化遗产。本土设计的策略：首先是保护历史遗产信息的原真性，其次是尽量不改变历史遗产本质已存在的环境特征，第三是要营造适合展示历史遗产文化信息的空间场所。

从重到轻的转变
Transition from heavy to light

江苏园博园主展馆及傲图格精选酒店 · MAIN PAVILION AND AUTOGRAPH COLLECTION IN JIANGSU GARDEN EXPO
设计 Design 2018~2021 · 竣工 Completion 2021

地点：江苏南京 · 建筑面积：52 374平方米
Location : Nanjing, Jiangsu · Floor Area : 52,374m^2

合作建筑师：关 飞、董元铮、付轶飞、刘亚东、毕懋阳、王德玲、张嘉树、郭一鸣
　　　　　　窦 强、刘佳凝、关 晖、邓笑欢、郑碧芳、时 红、宋旻斐、卫嘉音
Cooperative Architects : GUAN Fei, DONG Yuanzheng, FU Yifei, LIU Yadong, BI Maoyang, WANG Deling, ZHANG Jiashu, GUO Yiming
　　　　　　　　　　　DOU Qiang, LIU Jianing, GUAN Hui, DENG Xiaohuan, ZHENG Bifang, SHI Hong, SONG Mingfei, WEI Jiayin

策　略：保护、利用、对比、对话

摄影：张广源、李季、侯博文
Photographer: ZHANG Guangyuan, LI Ji, HOU Bowen

江苏园博园 9.13

江苏园博园主庁饰。
　尺子. 会议. 商业服务…… 2.5万m²
· 2020. 9月开园. 需要两年.
地点: 江宁区 汤山. 茅茂山麓. 距之句山民工改造.
山地. 西排毛方浅保留. 有主坑和时对.
年份: 抓工业遗兒保留与园艺厅二者结合并走. 我想
　　新巴洛克. 球之阵范围. 园间. 园管.
　园斗. → 园开. 园卸. 园塔. 园台.
　钢构件. 钢节点. 钢桁架. 钢栈道.

城市很美好风景……

2020.9月有感。背面。

十度：シ方位ぶあり、差がない。長さな比に7.5度。

路：也理由是日常生活，商方主不左方向。

角：四个建到处是马匹产生是冷害。

园林: → 园内、园林, 园外, 园界.

房屋内, 墙内外, 窗内外, 信息界限.

　　项目选址具有相当的独特性，以一座1970年代水泥厂的厂区遗址作为建设基地，以修复生态、织补城市功能、创造绿色美好城市的新型公共绿化空间为目标，尊重"生态"与"遗产"是我们在主展馆设计中秉持的重要价值观。

　　"轻介入"的设计策略使破败的工业废墟重生为绿意盎然的现代园艺展馆，为南京市民提供一座富有体验性和生态性的公共建筑，成为"永远盛开的南京花园"。"轻介入"，体现在"轻的结构"——采用装配式钢结构系统，实现环保和快速的建设过程；"轻的形象"——纤细的结构构件结合攀爬类垂直绿化，打造轻盈的建筑形象，与工业建筑的粗糙厚重产生对比；"轻的态度"——无边的绿色花园弥漫在工业遗址之上，使建筑消解自身形象，消失在自然之中。

　　"轻介入"的设计策略也使建筑具有更高的灵活性，以适应多种功能、多种类型空间的需求，同时兼顾展后改造和再利用。

The design takes the site of an abandoned private cement factory as the construction base. Restoring environment, repairing urban functions, creating a new public green space in the city and respect for "ecology" and "heritage" are the primary value orientation during the planning and architectural design process of the main pavilion.

The new building adopts the "light intervention" design strategy, cultivates the ruins into modern horticultural pavilions full of greenery. Light intervention, reflected in the "light structure" - the use of prefabricated steel structure system to achieve environmental protection and rapid construction process; "Light image" - slender structural components combined with climbing vertical greening to create a light architectural image, contrasting with the roughness and thickness of industrial buildings; "Light attitude" - boundless green gardens permeate the industrial site, allowing the building to dissolve its image and disappear from nature.

江苏园博园主展馆及傲图格精选酒店 · 2018—2021 · MAIN PAVILION AND AUTOGRAPH COLLECTION IN JIANGSU GARDEN EXPO

保护：对水泥厂遗址的核心厂房和生产设备进行精心加固和修复，结合活化后的观展流线，将生产线纳入新功能与新空间中，使其得到妥善保护和展示。

利用：水泥厂的多组筒仓改造为书店、酒店大堂等，原真的高耸圆筒得到合理利用，并成为新空间的有机组成部分，为新空间注入了特色。

对比：主展馆在保护和利用旧水泥厂的同时，刻意将加建部分采用钢结构，与既有工业建筑封闭厚重的混凝土砌体结构形成对比，让人能分清新建筑与历史遗迹的时空关系，同时体现了时代的技术变迁。

对话：主展馆的新建建筑，通过采用与既有工业遗产相似尺度的几何母题，形成新旧之间在节奏和尺度上呼应和对话。

剖轴测图

江苏园博园主展馆及傲图格精选酒店 · 2018—2021 · MAIN PAVILION AND AUTOGRAPH COLLECTION IN JIANGSU GARDEN EXPO

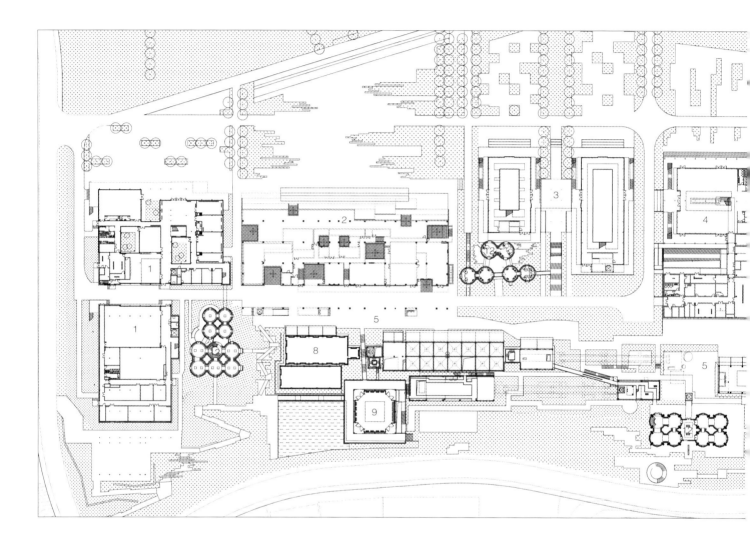

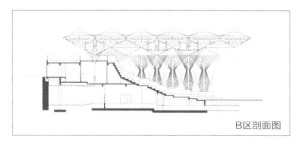

B区剖面图

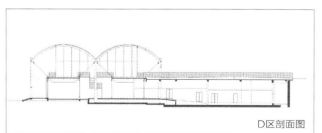

D区剖面图

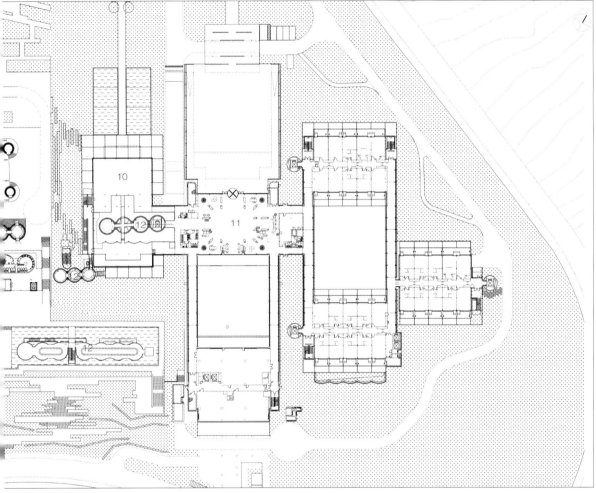

1. A区办公区
2. B区主入口广场
3. C区展厅
4. D区餐饮区
5. 商业街
6. 酒吧
7. F区商服区
8. 室外展场
9. E区展厅
10. 酒店公区
11. 酒店
12. 筒仓

首层平面图

A区剖面图

江苏园博园主展馆及傲图格精选酒店 · 2018—2021 · MAIN PAVILION AND AUTOGRAPH COLLECTION IN JIANGSU GARDEN EXPO

LAND-BASED RATIONALISM III   *105*

江苏园博园主展馆及傲图格精选酒店 · 2018—2021 · MAIN PAVILION AND AUTOGRAPH COLLECTION IN JIANGSU GARDEN EXPO

江苏园博园主展馆及傲图格精选酒店 · 2018—2021 · MAIN PAVILION AND AUTOGRAPH COLLECTION IN JIANGSU GARDEN EXPO

LAND-BASED RATIONALISM III *109*

江苏园博园主展馆及傲图格精选酒店 · 2018—2021 · MAIN PAVILION AND AUTOGRAPH COLLECTION IN JIANGSU GARDEN EXPO

江苏园博园主展馆及傲图格精选酒店 · 2018—2021 · MAIN PAVILION AND AUTOGRAPH COLLECTION IN JIANGSU GARDEN EXPO

江苏园博园主展馆及傲图格精选酒店 · 2018—2021 · MAIN PAVILION AND AUTOGRAPH COLLECTION IN JIANGSU GARDEN EXPO

江苏园博园主展馆及傲图格精选酒店 · 2018—2021 · MAIN PAVILION AND AUTOGRAPH COLLECTION IN JIANGSU GARDEN EXPO

江苏园博园主展馆及傲图格精选酒店 · 2018—2021 · MAIN PAVILION AND AUTOGRAPH COLLECTION IN JIANGSU GARDEN EXPO

设计随笔

对园博会意义的思考和对工业遗产的态度

江苏省园艺博览会由江苏省人民政府组织，于1999年在南京召开首届，此后两年举办一届。博览会作为重大的城市事件，最早可以追溯到18世纪末的欧洲，工业革命后激增的技术发明、工业产品以及旺盛的全球化贸易需求，催生了市民节庆一般的盛大展会。近年来，专门分享园艺技艺和园林文化、倡导绿色生活的园艺博览会，也伴随着中国社会生活的进步需求而在中国繁荣。

早期大博览会的主展馆，是专门为博览会设计和建造的临时性建筑。虽然有临时性，却又常常因为带有强烈的成就展示和宣传的目的，而成为当时建筑技术和艺术的风向标。从水晶宫到埃菲尔铁塔，都因其建筑材料（钢铁和玻璃）、结构（大跨度）的突破性在建筑史上占有一席之地。工业革命的爆发催生了博览会这一事件，也造就了博览建筑的工业基因：快速甚至是临时性建造，追求通用性、标准化的展览空间，以及在建筑材料和技术的革新性上寻求突破。可以说，博览建筑与现代工业所代表的效率和美学相伴相生。

工业遗产作为文化遗产中一种年轻的类型，大部分不具有文物建筑的价值和唯一性，但它作为具有代表性的生产空间，曾经大量存在，并塑造了某一时期的社会生产和生活。因此，不计代价地保护和维持原状不是目的，真正的目的是使其在留存核心信息的基础上得以再生、让旧空间能更好地容纳新功能和新生活。我们在对待工业遗产的态度上，对保护和更新的取舍度量，正基于此。

保护什么？保护厂区原始的场地标高和山地特色；保护建筑的天际线；保护主要建筑与场地的图底关系；保护具有典型工业风貌的建筑外观；保护生产线尤其是钢设备的完整性。

更新哪些？更新近人尺度的建筑界面，增加空间的开放性；更新楼电梯和步道连桥，增加建筑的可达性；更新结构系统以确保延长使用年限的安全性；更新外围护构造和机电系统以获得旧建筑物理性能的全面提升。

举几个具体的策略——1）将新建筑的体量和立面消解在更小的尺度中，高度也尽量压低到2~4层，以保持旧建筑的分量感，使旧建筑的天际线仍然统治着大地。改造更新后主展馆的剪影，没有改变这里旧日印象的轮廓。2）旧建筑整体以保留原有的立面为主，维持工业建筑的朴素形象，仅在底层打破封闭的格局，拆除外墙，置入轻型建构的装配式门廊和橱窗，形成开放和富有人情味的商业街道。

在如何对待既有建筑的外观风貌这个问题上，有个问题一度困扰我们：对于旧建筑，构成风貌印象的不可忽视的要素是外墙砂浆。立面上的风雨侵蚀甚或污染和霉变的痕迹，正是建筑师和很多体验者眼中难得的岁月质感和"废墟情调"。但是砂浆大面积空鼓、剥落的现实，使保护这种"情调"成为代价极为高昂的选择。基于前述的价值判断，我们选择根据现实情况铲除恶化的砂浆，顺应构造需求引入新的砂浆面层解决墙体机能性的问题，满足延长砌体寿命、保证砌体耐久度和安全性的刚性需求。针对新砂浆的质感和颜色，我们参与了大量的现场实验。为了增加功能性砂浆抹面的美感，采用了一种粗颗粒憎水砂浆，由工人进行自然笔触的手工抹灰，用微妙肌理修饰并不平直的墙面。最终，工业遗产的基本风貌与历史特征还是匹配的，同时又增加了一些更微妙的表情。

工业理性的启发与网格体系的建立

旧工业建筑的空间组织和场地利用，直接、朴素地表现出工业生产对效率的追求：有序、理性、直接、组织、安排、运转，转换为建筑语言则表现为线性布局、高差错落、空间最小化，成串的重复的基本几何形体（圆筒仓和方厂房），勾勒出电路板一样分明的图底关系。

主展馆的新建筑总建设量不小,但需要在复杂标高和旧建筑共同切割而成的三维畸零空间中有序地组织。在场地既有建筑线索的提示和工业理性的启发下,我们想到引入一个网格体系,作为新建筑和场地组织的辅助——一个4m×4m×4m的三维轴网,一个可依附的空间坐标。新空间和新景观可以在隐性空间坐标的控制下有序蔓延,新的建造成为均质的"底",旧的存在仍然是特异的"图"。

4m尺度的选择,来自于消解新建筑巨大体量的企图,和对场地道路、建筑层高、平面基本模数的兼容。4m网格控制场地上一切新的置入。首先引入垂直于台地高差方向11条林荫路,以相互平行的方式切割出大小不同的可建设区域,这种垂直切入为场地的漫游感知提供了新的路径和新的体验。新置入的建筑体量在林荫路之间的正方形网格上落位、重复、堆叠、虚实转换,形成了尺度相近的一组组建筑群体和院落。

4m边长的正方体成为新空间和体量感知的基本单位,与场地上4~5m直径的圆柱筒仓形成的既有线索具有相关性——同样的单纯几何空间、相似的串并组织关系、协调的模数与尺度。复杂的场地与新旧建筑彼此之间,具有清晰可读的关系,也能容纳微妙的联系。

轻钢结构体系与水泥厂体系的对比与对话

相比旧建筑的封闭、实体感,我们希望新建筑以一种消融和暧昧的状态存在。这种暧昧一方面是为了与工业建筑肯定的形式感造成反差,另外一方面,是从精神性上表达一种建筑师的期冀:消弭人类在这片场地上的再次建造与自然产生的对抗;融和人造空间与自然空间的隔绝和割裂。

因此,我们在功能体量的外围,将抽象的三维网格体系物质化——灰绿色的70mm截面方钢,沿着三维网格,组成销接的格构系统。格构系统附着在主体之外一个跨度的进深,在空间上,成为室内外空间之间的灰色过渡;在界面上,消解了明确的二维立面;在功能上,承载了阳台、绿植、雨篷、外廊等次级构件和附属功能。对人的体验来说,从室外的阳光风雨到室内由精密的空调设备营造出的恒温空间,存在着这样一方天地,既非绝对舒适,也非毫无庇护,这是人造空间向自然延伸的一种方式——以一种自身消融的态度。

当一个完整的、严密的几何系统,控制了一座建筑的平面、立面、结构体系,直至装饰和围护系统,它的设计与建造自然获得了效率与秩序。新建筑在物质层面上,是清晰、纤细、理性、精准的,而在人的体验上,是柔和、模糊的,带着风和光线的滤过。工业遗存的封闭、厚重、苍凉、粗犷,在新建筑的纯净、精致的对比之下,激发了一种新的美学体验。注6

圣贤之地的谦恭
Humility in the homeland of Confucius

曲阜鲁能JW万豪酒店 · JW MARRIOTT HOTEL QUFU
设计 Design 2007 · 竣工 Completion 2019

地点：山东曲阜 · 用地面积：80 900平方米 · 建筑面积：55 000平方米
Location : Qufu, Shandong · Site Area : 80,900m² · Floor Area : 55,000m²

合作建筑师：陆静、何佳、成心宁、杨茹、张男
Cooperative Architects : LU Jing, HE Jia, CHENG Xinning, YANG Ru, ZHANG Nan

　　曲阜鲁能JW万豪酒店位于山东曲阜明故城内，与孔庙仅一街之隔。设计中建筑布局与孔庙的关系以及与古城肌理的关系被放在首位，酒店规划遵循孔庙的轴线，参考孔府的院落形态，借鉴阙里宾舍的合院尺度，将公共区、客房区与孔庙、孔府分别对应。建筑形态借鉴传统元素，坡屋顶形态错落有致；酒店公共区屋顶举折明显，形成优美的曲线；客房区屋顶采用卷棚顶，轻巧舒适。公共空间汲取山东园林的特点，以方亭作为空间核心；客房以民居院落为原型，院墙和院门半围半透。外墙采用装饰多孔砖复合外墙技术，用绿色环保材料创造出古建丝缝墙效果。场地内保留大量古树，景观围墙内凹形成多处开放公共绿地。

The site of the hotel was just across the street from the renowned Confucius Temple, and the design was thus challenged by the requirements of respecting historic relics and accommodating the functions of a hotel into a courtyard-style layout. The axis of the whole project is parallel to that of the Temple of Confucius. Since the Temple consists of palatial buildings while the Mansions of Confucius are dominated by residences, to reflect such a contrast, the hotel's divided into a public area and a room area. In the public area, the raising of truss is adopted; while ridge-free roofs of the room area remain low-profile. Traditional building elements such as floral-pendant gates and spirit screens are applied in the buildings in a modern manner.

曲阜鲁能JW万豪酒店 · 2007—2019 · JW MARRIOTT HOTEL QUFU

1. 公共区
2. 宴会区
3. 客房区
4. 原有保留建筑
5. 孔庙
6. 孔府
7. 阙里宾舍

总平面图

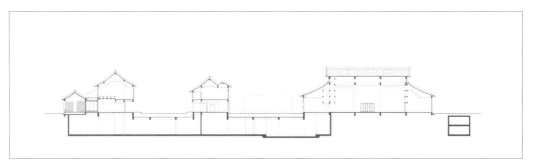

剖面图

LAND-BASED RATIONALISM III　123

曲阜鲁能JW万豪酒店 · 2007—2019 · JW MARRIOTT HOTEL QUFU

曲阜鲁能JW万豪酒店 · 2007—2019 · JW MARRIOTT HOTEL QUFU

协调：以尊重的心态读懂既有环境的典型语汇，使新旧能够相处友好，积极对话。

织补：梳理既有环境中各种城市要素，精心对接，巧妙补位，让城市因新的设计更完整。

重构：对既有城市闲置资源重新激活，积极利用，在对空间的重新组织和结构安排中呈现新的的价值。

容错：城市很难完善，之前的问题和错误绝不鲜见。直面这些问题，不随意大拆大改，积极地调整和改良，顺势而为，常常是有效的办法。

植入：在历史截取中植入有活力和特色的小型变异体，形成对比和特殊的体验性。

演变：城市的发展是个渐进的过程，设计应该描述这一进程，不保守不激进是一种不动声色的转变。

渐进：利用新旧语言的混搭、穿插，描述有机更新的城市生命体的复杂性，体现有机生长的动态性。

城市更新
URBAN RENEWAL

城市是有生命的综合体，在逐渐发展中不断进化。城市风貌是多元融合的，它不仅依赖于标志性建筑，还存在于城市街区公共空间以及大量的背景建筑之中。城市更新是一种织补，也是一种演进，要保护和利用并重，修补和创新并举，旧中出新，新中有旧。顺应和描述这种进化的状态是本土设计的策略。

向经典致敬
Salution to the classic works

武汉大学城市设计学院教学楼　SCHOOL OF URBAN DESIGN, WUHAN UNIVERSITY
设计 Design 2017 ・ 竣工 Completion 2021

地点：湖北武汉 ・ 用地面积：11 045平方米 ・ 建筑面积：13 280平方米
Location : Wuhan, Hubei ・ Site Area : 11,045m² ・ Floor Area : 13,280m²

合作建筑师：喻弢、胡水菁、曹洋、张笑彧、周益琳、金爽、伊拉莉亚、冯君
Cooperative Architects : YU Tao, HU Shuijing, CAO Yang, ZHANG Xiaoyu, ZHOU Yilin, JIN Shuang, Ilaria Bernardi, FENG

项目位于武汉大学珞珈山校园内，东、北两侧紧邻学生宿舍，南侧与狮子山隔路相望，场地紧凑狭长，高差变化复杂，周边植被茂盛，交通繁忙。作为城市规划建筑设计专业的教学楼，在满足面积、功能及辨识度的同时，更需提供高效灵活、具备设计教学示范作用、富于趣味的创新空间。

建筑顺应地势高低错落，教室外的挑台层层叠叠，密肋吊顶和深棕色断桥铝合金玻璃幕墙重现民国建筑庄重典雅的木构件意向，屋顶为覆盖着绿植缓坡的檐部，环绕布局的上人平台可眺望"樱顶"和东湖。北侧外墙披覆蓝绿相间的金属格栅，参数化设计的深浅色系分布以及疏密间隔既是对校园老建筑孔雀蓝色琉璃瓦的呼应，也避免了与对面学生宿舍间的视线干扰。

因场地限制又需设置尽量多教室的原因，形成了绘图教室短边对外的格局，设计通过各年级教室空间的连通，特别是在4.5m层高中穿插6m高的空间，使整个空间敞亮均匀，层层抬高的布局也十分利于空气流通。西侧教学区的设计系教室围绕两层高中庭组织，通高玻璃幕墙映入对面郁郁葱葱的山景。教室外皆设有层层平台，站在平台上伸手则可触及场地内阔叶树枝丫，将自然融入日常的教学空间。

Located at Luojiashan Campus of Wuhan University, the School of Urban Design is located on a narrow and uneven site bordering students' dormitory at its east and north, with Lion Hill at its south across the road. As a teaching building of urban planning and architectural design, the project aims to present places that are flexible, creative and fun.

Going along the landform, the building has a series of terraces outside the classrooms, with ribbed ceiling and dark-brown glazed walls resembling images of buildings in the Republican period. On the roof, cornices are covered in green, and walkable platforms allow people to overlook the cherry blossom and East lake. On the north façade, blue and green metal gratings echo with the peacock-blue colored tiles, while buffering direct visual interference with the students' dormitory.

To fit as many classrooms as possible in a limited area of space, the short sides of drawing classrooms are paralleled with the façade. 6-meter high spaces are inserted among 4.5-meter high spaces for various grades, balancing the brightness of spaces while boosting ventilation. Students can touch the leaves on terraces outside the classrooms, where nature is introduced into daily teaching activities.

武汉大学城市设计学院教学楼 · 2017—2021 · SCHOOL OF URBAN DESIGN, WUHAN UNIVERSITY

协调：消解体量融入山林，绿坡屋顶与琉璃呼应，土红色窗框与老建筑相仿，阶梯式空间如山坡步道。

场所：保住了大樟树，也留下了师生的"乡愁"。以树为中心营造了景观共享的前庭，如树般斜挑的柱子让出了更开放的首层，山和楼在这里相融。

织补：嵌入狭窄用地让空间不显得拥挤，架空体量让人流、气流都比较通畅，无需改变地形，顺坡组织不同标高的空间入口，前挑的阳台与后仰的山林相拥。

重构：顺应场地的条件，顺应环境的要素，顺应使用的逻辑，重新构建场所和空间并取得突破。

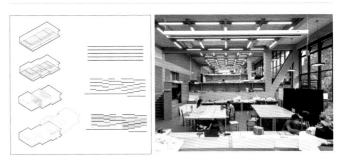

演变：对校园历史的尊重、对历史建筑的借鉴、对当代建筑功能的引导、对当代技术的应用，所有要素的真实呈现表现出一种演变的过程。

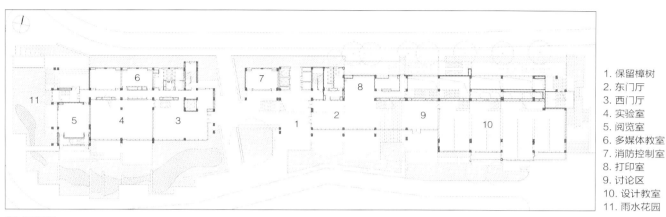

首层平面图

1. 保留樟树
2. 东门厅
3. 西门厅
4. 实验室
5. 阅览室
6. 多媒体教室
7. 消防控制室
8. 打印室
9. 讨论区
10. 设计教室
11. 雨水花园

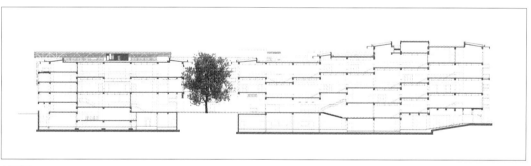

剖面图

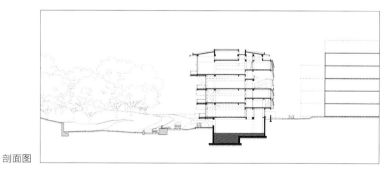

剖面图

武汉大学城市设计学院教学楼 · 2017—2021 · SCHOOL OF URBAN DESIGN, WUHAN UNIVERSITY

LAND-BASED RATIONALISM III *137*

武汉大学城市设计学院教学楼 · 2017—2021 · SCHOOL OF URBAN DESIGN, WUHAN UNIVERSITY

武汉大学城市设计学院教学楼 · 2017—2021 · SCHOOL OF URBAN DESIGN, WUHAN UNIVERSITY

设计随笔

我第一次进武汉大学还是40多年前研究生调研的时候。记得那次在绿林浓荫中前行，一座座古朴而优雅的历史建筑掩映其中，尤其印象深刻的是山坡上的学生宿舍楼，穿过大拱门，爬上层层的登山台阶，两侧就是一排排的宿舍楼，空间很有节奏也很有效率，一直爬到屋顶平台，能俯瞰美丽的校园和山景湖光连成一片。

再次进入武大就是八九年前了。当选院士之后，有幸在珞珈讲堂做过一次讲演，又趁机在校园中粗粗转了一下，举目都是形式各异的校园新建筑，布局也比较拥挤和散乱，似乎老校园的格局被打乱了，老校园建筑也退在树林之后，成了配角，似乎并不再是武大的主人了，这给我留下了不佳的印象。那次讲演后不久，我便接到武大基建处的电话，邀请我为城市设计学院做个新教学楼，并说规模不大、投资有限，问我有没有兴趣？我欣然应允，心中想着新的设计一定要向老校园致敬，绝不能自说自话彰显自己！

当我带着任务再次走进武大，当天晚上便沿着珞珈山林中的步道与助手边走边感受那种历史校园的气息。清早吃饭前爬上山顶，到正在维修中的老图书馆和教学楼去学习、观摹，从地形的利用到空间的组织以及许多细节的处理都很精到，并没有因为时代的局限、技术的陈旧而失去设计的价值。从山上下来，山脚处的一座破旧的教学楼就是城市设计学院。那时已是人去楼空，准备拆除。我们逐层踏勘，从底层直到屋顶。用地十分狭窄的教学楼被前后两排树夹在中间，楼内平面简单，原来是武汉水利电力学院附属中学的教学楼，只在一层有一玻璃大厅，听说是当年张在元先生任院长时改造出来的。窗外一棵大樟树遮天蔽日投下斑驳的树影，留下了一点设计的感觉。楼上层层外廊虽然单调，也给教室提供了看山望林的空间；屋顶虽然架空板破损严重，但可以遥望东湖的美景；外立面虽平淡无奇，但张在元先生细心涂上的粉绿色的横饰倒是让楼有几分特别的细腻，也似乎与老校园的蓝绿琉璃瓦有某种呼应。

用地窄且长，地势平且坡，树林近且密，屋顶宽且直。这次现场的观察和行走，让我很快找到了设计的思路，在回程的火车上我勾出了最初的草图。横向用足长度，教室面南一字排开；竖向顺坡为势，设计教室错台相通；剖面下小上大，层层出挑，让斜柱像树枝一样伸展；屋顶草坡勾边向蓝琉璃致敬，平台居中不仅可观湖，更可用于教学搭建的场地；以大樟树为轴，让底层架空连通前后道路；教师小间顺直在北侧，与教室隔层相连，降低了层高，增加了层数；学术报告厅利用地面起坡让地下和地上空间连续；连路角处原来的快递收发点也用架空、地台和上部棚架，形成有场所感的积极空间。北侧设计了轻巧的绿色横条金属格栅，既遮挡了与对面宿舍楼的视觉干扰，也算是对张在元先生留下的横条纹记忆的一种呼应。我的构思虽然来得快，解决问题也比较到位，但更重要的是我的助手和设计团队下了功夫，用BIM精心构建精细模型，"粗粮细作"，全程配合，用几近无装修的结构性和功能性要素表现建筑朴素的美和教学的价值，更表达出我们对武大经典校园的致敬！

山盘水绕
Winding mountains and rivers

重庆市规划展览馆 CHONGQING PLANNING EXHIBITION HALL
设计 Design 2020　竣工 Completion 2021

地点：重庆南岸　建筑面积：17 000平方米
Location : Nan'an, Chongqing　Floor Area: 17,000m^2

合作建筑师：景泉、徐元卿、李健爽、张翼南、颜冬、杜永亮
Cooperative Architects : JING Quan, XU Yuanqing, LI Jianshuang, ZHANG Yinan, YAN Dong, DU Yongliang

合作机构：中煤科工重庆设计研究院（集团）有限公司
Cooperative Organization : CCTEG Chongqing Engineering (Group) Co., Ltd.

策　略：改造、顺势、开放、整合、观景

摄影：张广源、郑勋
Photographer: ZHANG Guangyuan, ZHENG Xun

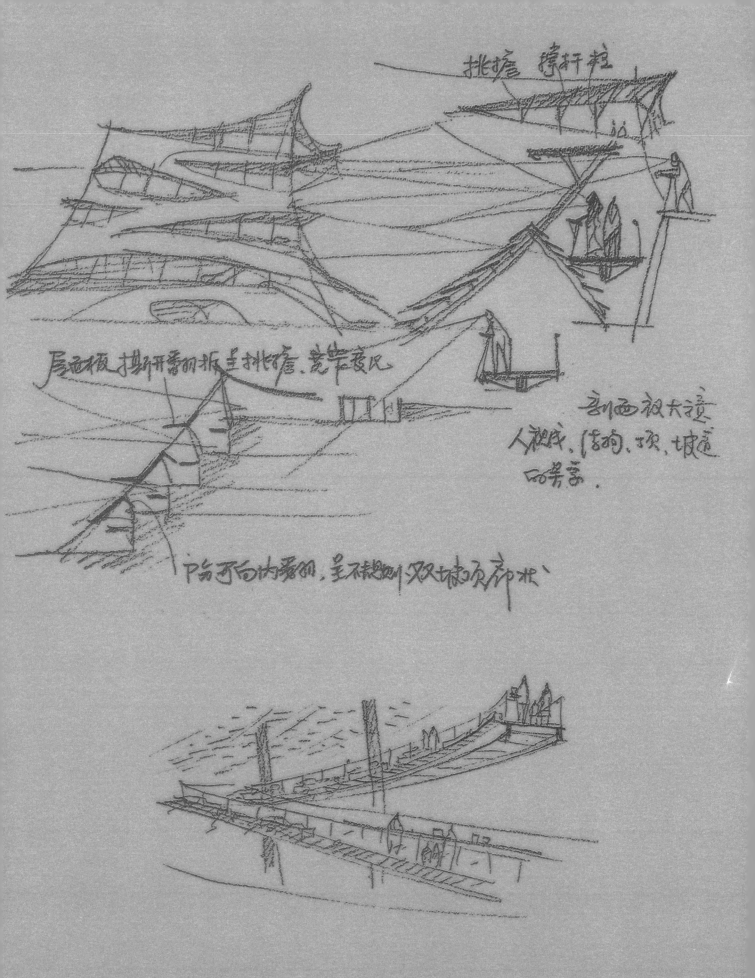

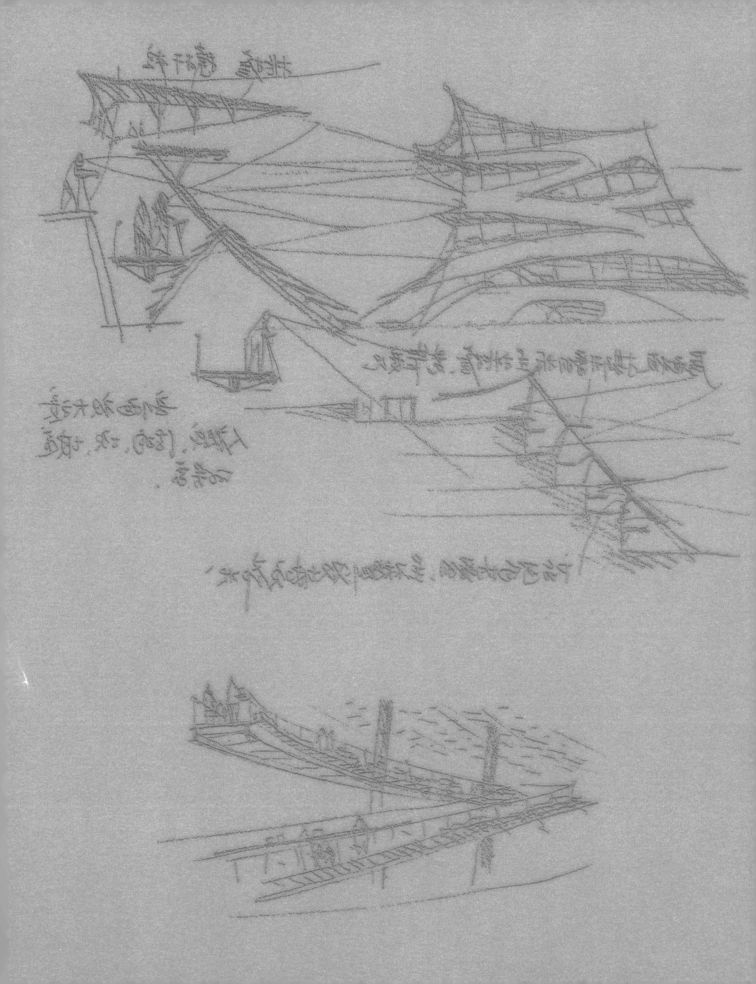

迁建后的重庆市规划展览馆地处长江与嘉陵江两江交汇处，与朝天门广场、江北嘴重庆大剧院遥相呼应，景观极佳。建筑由原有车库改造而成，以缝合山水环境、织补城市空间为出发点，探索了基于公共性、地域性的绿色营造方式。

改造基本保留了原建筑的弧形室外露台，在室外加建的"之字形"步道，从一层延伸至四层，以坡道、楼梯、平台共同构建了山地公共空间游览系统。不仅在纵向上打通了由滨江步道上山的流线，横向上也串联了弹子石广场周边的慢行系统，使建筑成为城市公共空间体系的一部分。

除了提供公共步行系统，加建部分也通过遮蔽步道的金属外幕墙产生了山城屋檐的意向，独具标志性。檐廊可在炎热的夏季有效降低灰空间温度，为公众提供舒适的观景空间，还能形成观赏两江四岸景色的写意景框。由步道、重檐提炼而成的建筑符号，更成为重庆城市形象的高度浓缩。

The relocated Chongqing Planning Exhibition Hall is at the crossing point of two rivers, with Chaotianmen Square and Chongqing Grand Theater, endowing the site with excellent views.

Renovated from a carport, the planning exhibition hall features sustainable construction that is land-based. Existing arc-shaped terraces were preserved, with newly built Z-shaped walkways extending from the 1st floor to the 4th floor, forming a multi-dimensional pedestrian system for the public spaces, making the building an integral part of the urban public space system and complementing the existing pedestrian system around the site.

Metal curtain walls cover the walkways, presenting features of the mountainous city. Doubling as eaves for the walkways, it cools the gray space and enhances thermal comfort for the public, who can view the spectacular rivers through a frame formed by walkways and eaves.

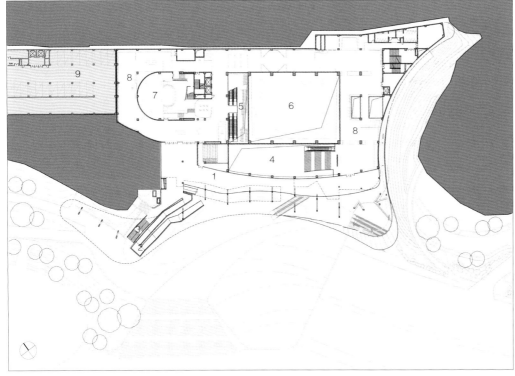

1. 室外平台
2. 室外坡道
3. 室外台阶
4. 序厅上空
5. "最重庆"中庭
6. 总规模型(不进人空间)
7. 飞行影院
8. 展厅
9. 弹子石车库(未改造部分)

二层平面图

剖面图

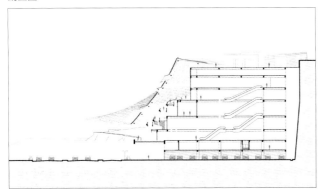

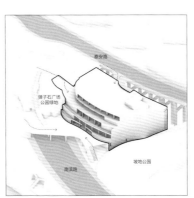

改造前体量分析图

改造后鸟瞰

改造前原状

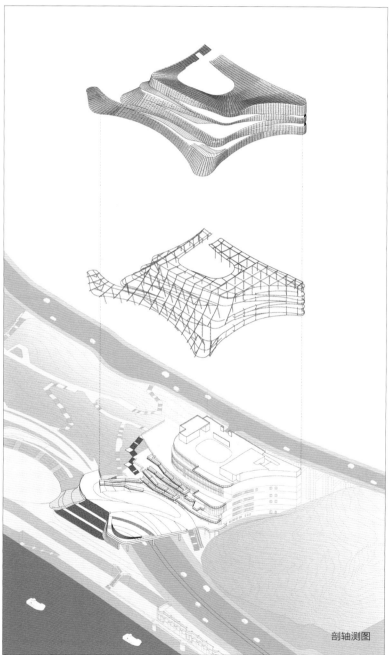

剖轴测图

重庆市规划展览馆 · 2020—2021 · CHONGQING PLANNING EXHIBITION HALL

让城市成为开放的庭园
Making urban plaza an inviting place for citizens

昆山大戏院 · KUNSHAN GRAND THEATER
设计 Design 2011 · 竣工 Completion 2017

地点：江苏昆山 · 用地面积：20 726平方米 · 建筑面积：50 553平方米
Location : Kunshan, Jiangsu · Site Area : 20,726m^2 · Floor Area : 50,553m^2

合作建筑师：刘恒、叶水清、梁世杰、Aurelien Chen
Cooperative Architects: LIU Heng, YE Shuiqing, LIANG Shijie, Aurelien Chen

策　略：开放、连接、协调、场所

摄影：张广源、Aurelien Chen
Photographer: ZHANG Guangyuan, Aurelien Chen

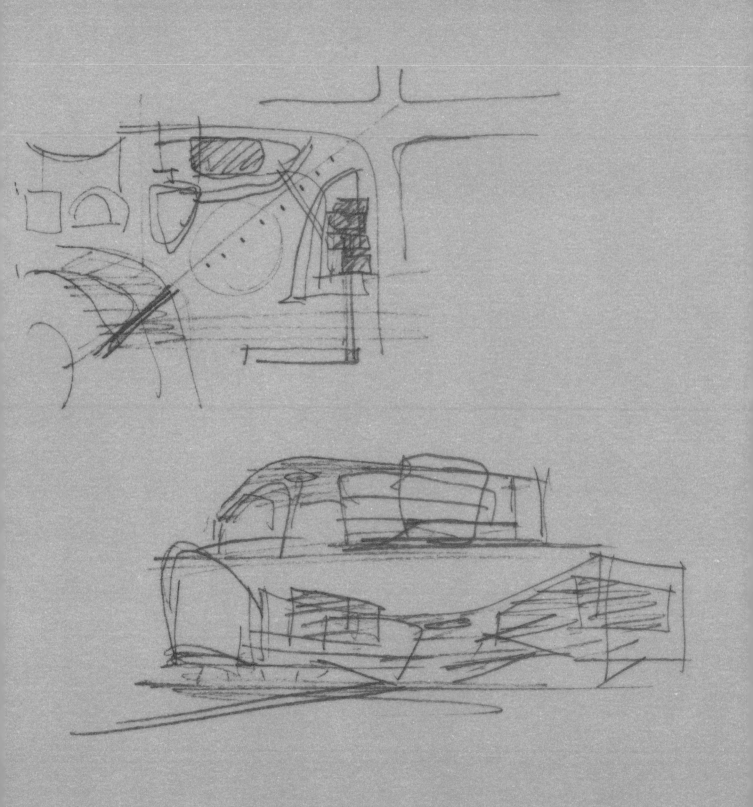

昆山大戏院位于昆山市中心城区的市民文化广场地块，在原有的昆山大戏院位置进行重建，并与图书馆、游泳健身中心及体育公园共同组成一个城市公共空间。通向体育公园的人流主要轴线从建设场地中间斜向穿过，将建筑分成东西两组，之间以流动形态的大屋盖相连，面对城市街角形成恢宏的入口广场。外围延续周边街区形成连续城市界面，内向则以层层错落的室外平台，结合不同部位的楼梯设计，形成连续贯通的围合空间。西北角的大台阶将图书馆前广场的人流引导至建筑内部；东南侧设置连廊与茶室、游泳馆及体育公园相连。三角形不锈钢顶棚与广场地面浑然一体，白天反射地面和行人，形成入口的点睛之笔；夜晚变幻的LED内透光则使这里成为吸引市民驻足、小憩、集会的场所。剧场休息厅内顶棚光纤灯疏密有致，不同颜色、尺寸的彩铝管墙面在光线作用下形成柔婉的"水袖"意向。

As a reconstruction project of the former Kunshan Theater, the porject has become a culture complex consists of a theater, a library, a swimming pool and a sports park. Curves, as a prominent feature of the building, have not only set the theme for the interior space, but also defined the feature of the building's exterior platforms. The artistry of the building is fully represented through the combination of decoration and illumination. The exterior wall of the theater lounge is decorated with red aluminum tubes, resembling the iconic fluttering sleeves in Kunqu opera, and the walls of the movie theater have a finish of Malay paint coated with stainless steel net. Reflections of the urban space, varying from day and night, are introduced into the building by the stainless steel mirror units under the huge roof.

昆山大戏院・2011—2017・KUNSHAN GRAND THEATER

1. 商铺
2. 大厅
3. 精品百货
4. 库房机房
5. 开架百货
6. 影院门厅
7. 专卖店

首层平面图

剖面图

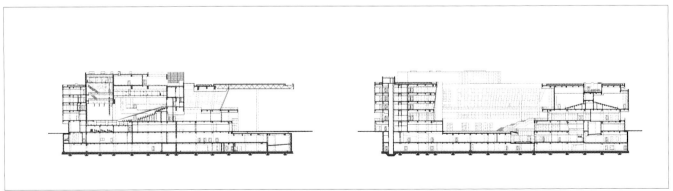

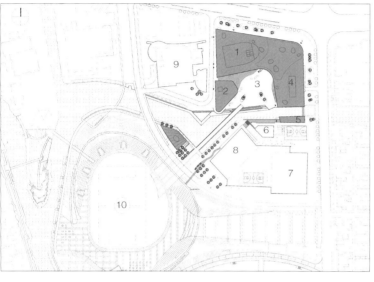

1. 多功能剧院
2. 电子阅览室
3. 中心广场
4. 电影院
5. 茶室
6. 室外游泳池
7. 篮球馆
8. 游泳馆
9. 图书馆
10. 体育公园

总平面图

改造前原状

昆山大戏院 · 2011—2017 · KUNSHAN GRAND THEATER

从烟厂到文化聚落
From tobacco factory to cultural settlements

宝鸡文化艺术中心　BAOJI CULTURE AND ART CENTRE

设计 Design 2014 ・ 竣工 Completion 2019

地点：陕西宝鸡 ・ 用地面积：20 726平方米 ・ 建筑面积：101 260平方米
Location : Baoji, Shaanxi ・ Site Area : 20,726m^2 ・ Floor Area : 101,260m^2

合作建筑师：时红、王可尧、刘洋、陈梦津、宋旻斐、冯君
Cooperative Architects : SHI Hong, Wang Keyao, LIU Yang, CHEN Mengjin, SONG Minfei, FENG Jun

项目地处宝鸡老城中心、金陵河和渭河交汇口，用地呈不规则形，南北约500m，东西约200m。这里曾是宝鸡市卷烟厂厂区，规模大，建筑保存完整。改造后园区内拥有五栋公共建筑——音乐厅、科技馆、群众艺术馆、图书馆和青少年活动中心，成为城市的会客厅、当地家庭休憩放松的场所。设计以保留体量最大的"C形"老联合厂房和结构完好牢固的高层宿舍楼为前提，利用连续的"几"字形体量将它们串联起来形成一个整体，"新"与"旧"相互连接，高低错落、曲折连续的曲面屋面使建筑群形成整体统一的沿街、沿河立面，而连续曲折的体量围合成的中小尺度空间提供了丰富而亲切的活动场域。外墙材料选用类钛金属色泽的金属板和金属格栅，以参数化方式在保证采光通风的同时，为不同角度的立面带来了柔和浪漫的变化。端部昂然高起的塔楼是建筑群的收束，也是周边城市环境的制高点，成为城市中心区的新地标。

Located at an irregular-shaped site at the center of the old city of Baoji, the art center renovated from a large well-preserved cigarette factory consists of a concert hall, a science museum, a public art center, a library and a youth center, serving as a city lounge and recreational center for local families. The largest C-shaped workshop in the plant and a high-rise dormitory building with good-conditioned structures are preserved and then connected with continuous zigzagged volumes. Curved roof has integrated various buildings into a whole, presenting unified facades both along the street and the river, with medium and small scaled spaces accommodating fun activities. Metal panels and gratings with titanium-like colors add variations to the façade while guaranteeing daylighting and ventilation. The tower rising at one end of the center dominates the height as a new landmark for the area.

宝鸡文化艺术中心 · 2014—2019 · BAOJI CULTURE AND ART CENTRE

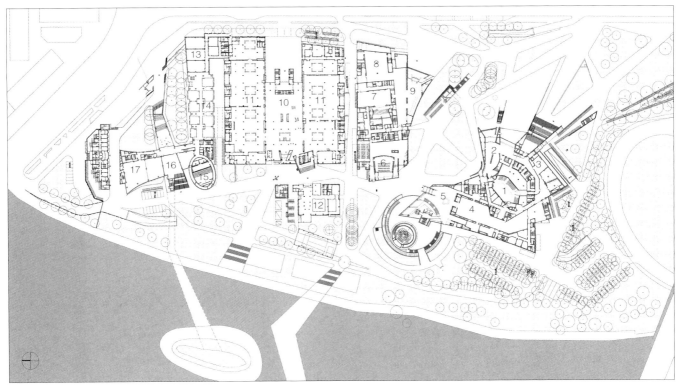

首层平面图
1. 音乐厅
2. 音乐厅门厅
3. 排练厅
4. 科技馆
5. 科技馆门厅
6. 科技馆4D影厅
7. 群艺馆大排演厅
8. 群艺馆美术展厅
9. 群艺馆门厅
10. 图书馆门厅
11. 图书馆阅览区
12. 书吧
13. 舞蹈教室
14. 美术教室
15. 多功能排演厅
16. 合唱教室
17. 运动类教室
18. 琴房
19. 球幕影院

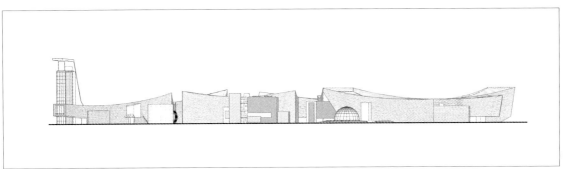

立面图

改造前原状

宝鸡文化艺术中心 · 2014—2019 · BAOJI CULTURE AND ART CENTRE

宝鸡文化艺术中心 · 2014—2019 · BAOJI CULTURE AND ART CENTRE

宝鸡文化艺术中心 · 2014—2019 · BAOJI CULTURE AND ART CENTRE

知识的累积
Accumulation of knowledge

中国大百科全书出版社办公楼改造　RENOVATION OF ENCYCLOPEDIA OF CHINA PUBLISHING HOUSE
设计 Design 2013　竣工 Completion 2016

地点：北京西城　建筑面积：21 312平方米
Location : Xicheng, Beijing　Floor Area : 21,312m^2

合作建筑师：吴斌、辛钰、范国杰、杨帆
Cooperative Architects : WU Bin, XIN Yu, FAN Guojie, YANG Fan

策　略：织补、重构、隐喻

摄影：张广源
Photographer: ZHANG Guangyuan

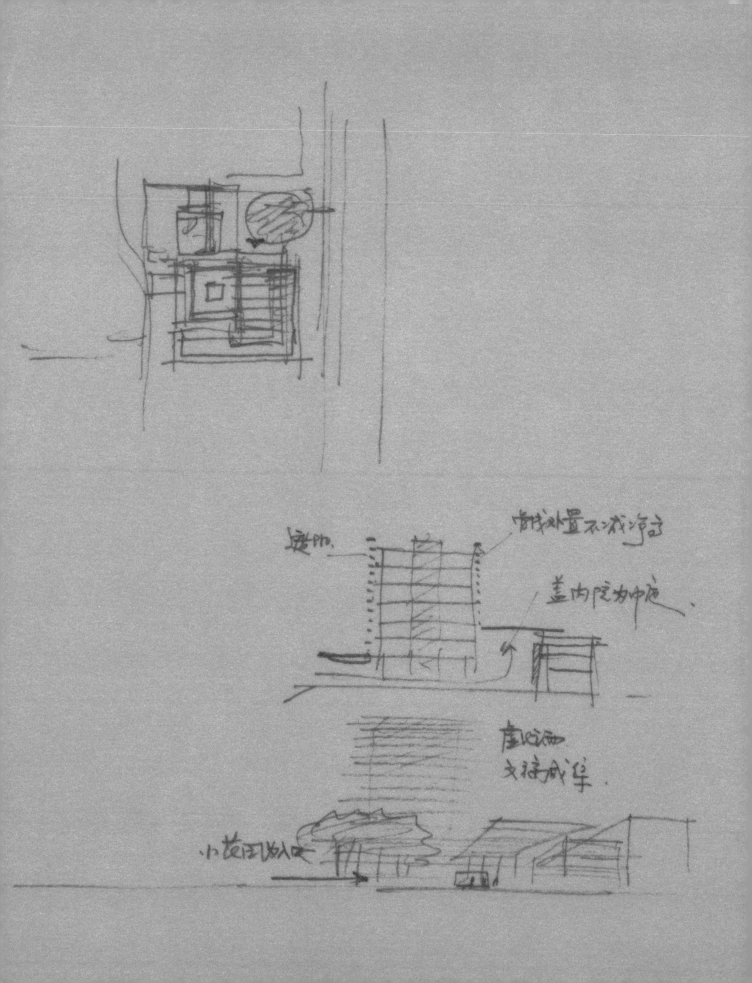

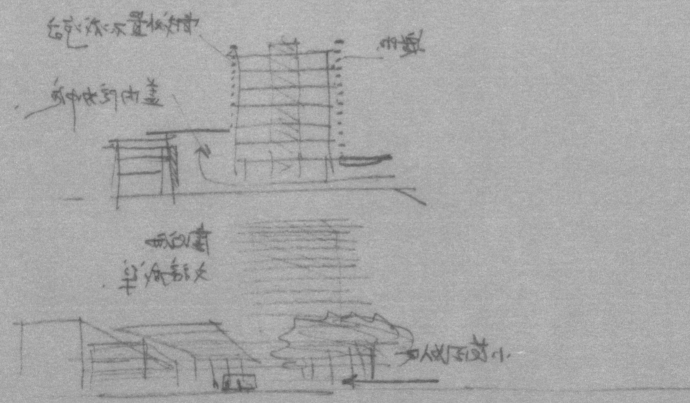

　　中国大百科全书大厦位于北京西二环路外，原有建筑于1987年建成，入口从西侧内街进入，道路狭窄，交通不便，小开间办公空间低矮局促。主楼和东侧裙房围合而成的内部庭院未得到充分利用，毗邻二环路的良好景观和形象价值也未得到有效挖掘。改造设计通过一体化设计的核心理念，将功能、管线、空间、外立面、文化品质等问题一并解决。建筑入口调整至二环一侧，让建筑回归主要交通路线，为原有内院覆盖玻璃屋顶成为室内中庭，兼具展览、接待、礼仪功能。为了释放内部空间提供开敞办公，打破传统方式，将主要设备管线沿外墙布置，并利用窗下墙空间进行整合。在外立面上用金属铝板将其包裹起来，铝板同时也起到遮阳作用，镂空格栅线条轻盈典雅，呈现出书页般的细腻和精致感，重新焕发出该建筑独特的文化气质。

The office building of Encyclopedia of China Publishing House is located along the West 2nd Ring Road of Beijing, but it could only be access from an interior street to its west, and a courtyard, as well as landscape of the West 2nd Ring Road was not effectively utilized. The renovation design, with integrated design as its core concept, solved a series of problems covering functions, pipelines, spaces, facades, cultural quality. The design relocated the building's entrance to establish direct access to the 2nd Ring Road, and the former courtyard has been covered with a glass roof, so that it has been transformed into an interior atrium for exhibition and reception. Main pipelines are located along exterior walls, integrated behind walls under the windows and covered by hollow-carved aluminum panels serving as sun shades while presenting the building's elegance.

中国大百科全书出版社办公楼改造 · 2013—2016 · RENOVATION OF ENCYCLOPEDIA OF CHINA PUBLISHING HOUSE

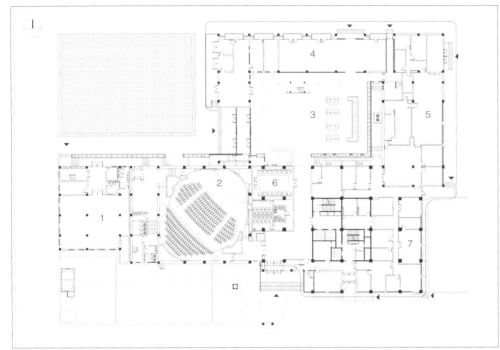

1. 大厅
2. 报告厅
3. 中庭
4. 展示厅
5. 营业厅
6. 贵宾休息室
7. 办公

首层平面图

剖面图　　　　　　　　　　　改造前原状

中国大百科全书出版社办公楼改造 · 2013—2016 · RENOVATION OF ENCYCLOPEDIA OF CHINA PUBLISHING HOUSE

关于城市之一：城市设计的维度与视角
ABOUT CITY I: DIMENSIONS AND PERSPECTIVES OF URBAN DESIGN

当前我国城镇化建设进入一个关键的转型期，从之前快速扩张式的发展向城市生态修复、品质提升、特色营造等方向转变。这几年党中央国务院对城市建设作出了一系列重要而具体的指示，住房和城乡建设部也把城市设计提高到城市管控的法定程序。许多大中城市的政府领导都在大抓"城市双修"和城市设计，而雄安新区以及多个国家级新区也为建设领域搭建了广阔的平台，提出了"千年大计，一张蓝图干到底"的口号，我国城乡建设的确迎来一个新时代！

在这样的大背景下，这两年城市设计也成为行业内热议的话题。比如城市设计与上位规划和下位建筑设计到底如何划定边界？城市设计到底包含哪些内容？是不是就是风貌设计、形象设计？城市设计导则到底规定到什么程度？如何用导则作为依据去审查建筑方案？能否作为土地招拍挂的必备条件？城市设计到底应该由规划师来做，还是建筑师来做？抑或景观设计师、市政工程设计师的工作是否也应统一纳入城市设计的范围来？

我认为城市设计可以有不同的维度、不同的层级和不同的视角。

鸟瞰视角

一般来说，提到城市大家都喜欢从空中开始看——一张规划图、一个城市模型、一幅鸟瞰图——这是反映城市整体格局和风貌的视角，当然是最重要的，也是各地领导最关心的，也是规划师们最要费神去想象的。为什么费神呢？因为如果是一个新区，他们并不可能知道这些建筑是什么？谁来投资？会不会是这个样子？只好先臆造一个图景，让领导满意，否则城市设计就很难被认可。其实大家都知道这种图景很难马上实现，实现时也很难不变样。如果是对现有城市来说，要达到鸟瞰图上的效果就更不容易，要么"穿衣戴帽"，要么大拆大改，其操作性的难度之大可想而知。也确实碰到几次政府领导问我：咱这个城市太乱了，是否您能定几个色，几个月内我让业主刷上去，城市不就有特色了么？凡此情况，我都很紧张，绝不敢贸然答应选什么色。因为我常见一些乡村就这么被刷新过，那种虚假和廉价的新面貌除了应对某些政绩活动，几乎毫无价值，钱花了但不会有什么好结果，更别说一个城市了，实在有些可怕。所以说，大家最常表现的、最期待和关注的城市鸟瞰视角的城市设计是很难实现的，越画得具体，越不真实，越没用。当然这并不是说这种层级的城市设计就没有用，而是在这个层次上更应关注城市格局的特色，比如城市规划格局与自然山水环境的关系，城市路网是否反映地形地貌特点，城市新区尺度、密度、高度与老城肌理的空间过渡关系，城市公共空间和生态绿化体系的关系等。这些设计意象对形成城市特色是十分重要的，也是能在规划审批中加以控制和引导的。而我在工作中也遇到不少在有山水特色的环境中粗放规划，将毫无特色的方格路网罩在有些丰富肌理的地形上，失去了创造城市空间特色的机会，很可惜。当我们碰到这种规划还未实施的时候，都会试着与政府商量，可否在建筑设计之前先做一个城市设计，调整上位规划格局，不仅能将有特色的山水环境保护下来，也使城市的空间有了特色。实在说，我虽然是建筑师，但我认为城市的特色比建筑的特色更重要。一旦错过了这种机会，建筑做得再好看些又能怎么样？而一旦城市格局有了大特色，房子好一点差一点也影响不大，许多国内外城市的案例都能说明这个问题。我们在福建南安市和四川仁寿县的两个项目中，通过城市设计，修改了原有规划，"救"下了一片山水环境，也营造了有特色的街区和建筑，心里就像做了件善事儿一样特别踏实，很有成就感。此为题一。

城市织补

如果我们并不依赖实现起来颇费时间和投资的大尺度的城市设计，那么提升城市品质亟需做的就是城市织补。在过往粗放式的城市建设中，许多建筑之间不协调，许多公共空间不连续，许多建筑和绿化用地碎片化，单靠嵌入新项目很难解决系统性的问题，因此城市设计在此关注的重点是城市织补。从点到线，从线到面，从既有建筑到新建项目，从建筑到步行空间、到公共广场和花园景观，从地上到空中、再到地下，城市设计几乎要触及城市空间中各个层面。而设计也是跨界的大设计，要打破城市建设中的条块分隔，开展各专业之间的积极合作，只有这样才能最终交给市民一个无缝衔接的、完整和谐的城市环境。应该指出，对城市存量的规划手段和思路与城市增量的方法是不一样的，用传统的规划套路处理城市织补的问题是完全行不通的，也下不去手。我们也不能再为了拉通城市街道网格去粗暴地裁切高密度的建成区，而为了历史街区尺度和肌理的完整保护，扩宽小街胡同、建筑退线、密度指标、绿地率等一系列规划管控手段也不再适用。所以用城市设计的视角和方法去做，采用"陪伴式"的设计服务是城市织补较有效的手段。近期听说北京规划和自然资源委员会在策划推动一千名（可能有点多）设计师负责一千条胡同的责任建筑师的服务模式，这是十分值得期待的。我们在北京天桥一块犬牙交错的小地块上设计的一个天桥传统文化传承中心的小项目和前门大栅栏 H 地块项目就是在这方面积极的尝试，虽然难度大，磨合时间长，但我认为方向是对的，项目也在推进中。此为题二。

城市街道

如果从城市空间的构成来讲，街道几乎是人人天天会用的最重要的公共空间。我提出"城市设计应该从脚下做起"，就是要关注城市街道设计，尤其是步行空间的品质。可是每天我上下班走过的街道，也就是住房和城乡建设部门前的三里河路怎么样呢？设计是按20世纪80年代流行的三块板式的断面，中间是上下四车道，然后是绿化（并不太绿）隔离带，外侧是非机动车道，现在基本上是停车带，再外侧是行道树，再外侧才是人行道，人行道再外侧是市政绿地，再外侧是建筑用地，这里多是政府机关的前广场或停车空间。这样的道路断面至今仍是城市规划中常用的经典范式，似乎没有什么改变和提升的必要，但实际上品质不高且问题不少。比如，街上随意设栏杆问题，主路中间设栏杆，据说为了防止人横穿马路和汽车随意掉头；公交车站设栏杆，据说为了让人排队；自行车道上设栏杆据说是这段不让停汽车；人行道和自行车道之间设栏杆据说是不让人横穿自行车道；人行道和绿化带之间设栏杆据说是不让人踩踏绿地；政府机关和人行道之间设栏杆是怕闲人进入影响安全。可见所有栏杆都是用来限制人的行为的，又怎么体现以人为本呢？还有的路段占路施工，人行道宽度被压缩到不足1m，还有的共享单车堆放在路边，行人只好绕开或躲闪前行，还有花池高台既不美观又碍手碍脚，还有的地方铺装简陋、常修常坏，别说景观设计、街道座椅和陈设，更别说两侧建筑对街道的开放度了……但是我想这些事儿要提升改造不难吧？比如能否用标识线、标识牌代替栏杆告知路侧停车的规定？能否用绿草绿篱代替栏杆来保护人的安全？能否用更多的树林代替草坪让人可以进入，又不必担心被踩坏？能否用临时的管理措施来防止突发事件而取消缺乏善意的铁栅栏？能否在街边为老年人设一些座椅？能否让街边商场有外摆的餐饮平台让人们可以在街上坐下来？但实际了解下来却并不简单，一个断面的管理权属分为多个部分，各部门有自己的规范规定、有自己的设计权限、有自己的建设主体，分界清晰，不能逾越，这就是为什么栏杆各做各的，地面各铺各的，景观和街道陈设各管各的，凑在一个断面上，不好用也不好看，行人感到处处被管，而不是处处有尊严地被服务。说实在的，这还是北京比较好的大街呢，更别说其他的背街小巷了。与国外城市相比，我们许多城市的街道品质不高，是造成城市印象不佳、文明程度落后的主要原因之一。因此要提升城市品质，就要从基础做起，从脚下做起，从小事和细节做起；而要做到这点，就要允许跨界设计、统一建设、综合管理；而城市设计的主要任务之一就是要跨越条块分隔，把街道当成一个整体的空间来设计，比起其他类型的城市设计任务，这是投资小、见效快、市民获得感强的大好事，为什么不优先做起来呢？2017年9月在王府井商业街上的绿池和坐凳系列是个快题设计，10天就搭建完成，得到了各方面的好评。后来我们又在无业主、无立项的情况下，利用课题研究做了一段三里河路步行空间的改造设想，也希望推荐给相关部门，争取能够实施。此为题三。

城市密度

我们在观察和体验城市时总觉得城市密度是个问题。我们的历史古城虽然是以低矮的平房为主构成的，但均质的高密度使其形成一个致密的整体。国外的历史城市亦然，五六层的楼房一栋栋比肩相靠形成了完整的街墙，石块铺就的小街尺度宜人，透过一个个门洞你可窥见大大小小的内院花团锦簇，优雅精美；即便是美国的现代都市中心，一栋栋高楼密集排列，楼下总有丰富的广场、花园和向城市开放的"灰"空间。应该说，高密度的城市尺度更宜人是不争的事实，但是我们在中华人民共和国成立以来沿用的规划方法却一直是以城市机动车道为主导的，根据预测的交通量去画路网、定路宽，形成的空间尺度当然就不太宜人，要想创造步行空间的应有尺度只能在保留的传统商业区中或建设用地内结合某些商业步行街设计去解决。另外，我们严格执行了半个多世纪的日照间距标准也是城市失去密度的重要原因。不管城里城外，不管中心区用地多么紧张，只要是居住类建筑几乎都要保证日照标准的硬性要求，所以许多原本历史上十分完整有序的城市格局在改扩建中变成无序和混乱的群簇状，但实际上这却是经过精心的日照计算而得到的结果，十分无奈和尴尬。还有诸如一定要在地块内设消防环路的不合理规定、建筑退线做地块内小市政管网使之与大市政分开的规定、各个建筑用地对绿地率指标的规定，凡此种种，就形成了谁也不担责任的松散无序的城市景象，的确今不如昔！那么今天我们对城市的织补，对建成区存量空间的改造和利用可否逐渐（肯定不能一下子）扭转这个局面，解决而不是继续恶化这个问题呢？我想这是设计的主要设计和研究方向。我认为对城市密度的研究可能会发掘建成区的空间资源潜力，在织补城市中大量存在的超尺度不协调的"剩余"空间时，可以增加不少为城市社区服务的小建筑。这些连续的多层小建筑就会形成宜人尺度的街墙界面，构建丰富的城市公共空间场所。而这个系统性的梳理和设计工作显然是难以通过单体项目设计来解决的。应该说明的是，增加密度并非是单纯为了在存量土地上提高开发强度，而是一种积极的置换策略，将原本被拆散的城市空间织补起来的同时，也可以开辟出尺度不同的小公园、小绿地、小广场，这比每个项目地块中勉强达标的绿地空间要积极得多、开放得多。所以通过城市设计可以把相关地块的控规绿地指标具体落地到一个合适的位置，让建设用地更加有效地利用，让置换出的指标合成一处，形成更有使用价值的城市公共空间。我想这会是城市设计中十分有意义的工作。此为题四。

风貌整治

近来我们还常碰到一类城市设计工作叫风貌整治。其实对已建成的沿街建筑拆违建、拆广告是一种管理行为，与设计似乎关系不大。但领导一看拆完了风貌还不好，是因为城市建筑立面不好看，所以希望用城市设计去指导整改立面；有的地方着急赶个什么大活动，干脆直接涂料画立面，粗糙地"化妆"，结果还不如那些貌似零乱破旧的立面所表达的真实生活状态和岁月留下的痕迹更好呢，这种花钱不落好的蠢事在城市更新中屡屡发生。但城市风貌确实需要设计，对既有建筑将来的改造也应有一种指导和建议。我主张结合都市生态化、公园化的发展大方向，对既有建筑进行立体绿化和垂直绿化的引导，如沿街住宅窗台上加装花槽花架；办公建筑加多层次的平台绿化和立面绿化；商场建筑更可以利用宽大的屋顶和大面积实墙做绿化种植，甚至适当增加立面上的步行外廊外梯，使立面空间化和开放化，让室内商业人流可以利用这种系统增加与城市空间的互动性，还可以结合节能的要求增加一些遮阳系统。我认为在规划和绿化管理上应给予相应的鼓励政策，立体绿化应纳入绿地指标进行折算，开辟城市公共空间应给予容积率补偿和奖励。这种在国外实行了很多年的管理政策行之有效，为什么不能应用到我们的城市中，以此来调动业主的积极性，主动改造自己的建筑，共建和共享生态花园城市？因此我认为城市设计一方面要有大思路、大智慧，要有引导城市空间转型的前瞻性，但又要有政策的支持和配合，才能逐渐实施，进入到有机的更新进程中去。那种寄希望于政府大投入、大包大揽的形象再造工程除非有特殊的大事件推动，多数是行不通的，也是不可持续的。因此城市设计的技术性和政策性同样重要。此为题五。

城市问题千头万绪，这五题并不全面，也不一定准确，但能说明一点，城市设计范围广泛，不能用一种标准、一种技术路径去解决，应该多维度、多层次、多视角地去研究、去观察、去体验、去设计，这是城市的需要，也是城市设计的魅力所在。规划师、建筑师、景观设计师、市政工程师、产品设计师、公共艺术策展人和艺术家都会找到自己的位置，发挥应有的作用。城市是大家的事儿！注7

关于城市之二：城市风貌的特色营造
ABOUT CITY II: TO CREATE CHARACTERISTICS FOR URBAN FEATURES

经过三十多年的快速城镇化，我们的城市日新月异，旧貌换了新颜。当我们为一派现代化景象而自豪的时候，当我们的经济发展指标也因为大规模建设而一路坚挺、高歌猛进、轰轰烈烈之后，我们忽然发现了一个大问题：在我们手中建设起来的城市如此雷同，如此乏味，城镇原来的风貌特色早已消失，因为代表城镇特色历史的那些老房子、老街、老树、老遗存早已在城市发展中被当作陈旧而丑陋的垃圾清除掉了；而在匆忙中建设起来的那些象征现代化的东西又好像是被孩子玩腻了的玩具很快就失去了吸引力，难以承担起人们寄托希望的城市标志性的角色，于是人们在冷静下来之后陷入了一种无奈和迷茫。

2012年党的十八大以来，中央的批示和社会广泛的关注让地方政府颇有压力，都在纷纷想办法打造城市风貌特色，试图在几年的时间内能够见成效。他们有的希望重新确定一种建筑风格，有的想从统一城市色彩入手，有的希望再造仿古一条街，也有的要打造一条滨水风光带……

应该说，这种城市建设转变总的方向肯定是好的、是积极的，以上这些愿望和做法也是可行的。但我觉得应该注意的是那种以往常见的急功近利的态度，因为实际上绝大多数城市的特色应该是很多年沉淀下来的，而不应是短期内"化妆整容"出来的；那些真正有价值的特色也绝不仅仅依赖城市外在的形态，而更有赖于它的历史和文化的内涵；那些真正令人感动和尊重的城市，也不仅仅在于它有值得炫耀的几处亮点，更在于那些无处不在的精致而友善的环境细节；当然还有文明、优雅的城市生活，更重要的是自然和谐的生态环境。显然，所有这些都绝非一朝一夕、三年五载可以成就的，但每一个正确的决策、每一项不多的投资、每一次真诚的努力都是十分重要的进步。

这就要求我们的政府彻底改变那种"一届一转向"、前后不持续的恶习；改变那种抓大放小、不注重细节的毛病；改变那种只重硬件建设，不注重管理运营的短板。总之，在我们今天的体制下，城市管理者的作风转变和价值取向对城市特色的形成影响很大，换句话说，今后城市的风貌特色也一定会反映出管理决策者的水平。

人们都说城市是个有生命的机体，它在不断地生长和完善，在过往的成长过程中有辉煌也有衰落，有新生也有陈旧，有亮丽也有丑陋。所有这些都会在城市的机体中留下痕迹，都是形成城市特色的资源，如何注意巧妙地利用、保护和有机更新，避免用破坏、遮掩和粉饰的方式是十分重要的，因为真实的、多元的、并不完美的城市远比一个经过刻意打造和装扮出来的虚伪城市可爱得多，这也是我比较反对那种片面强调打造城市风貌特色的理由。但是对并非都专业的城市管理者们来说，如何找到适合自己城市的、具有可操作性的做法，也是一个亟待解决的问题。近几年，越来越多的学者、专家和规划设计单位都在开展相关的研究和实践，取得了一些积极的成果，国际上的许多优秀案例也为我们提供了很好的参考和学习的示范，但这个过程中切忌照搬照抄，也要避免模式化和套路，因为我们追求的是不同地域、不同城市的不同特色。

覆土：以屋顶覆土的方式让建筑与大地风景浑然一体。

嵌入：以体量嵌入大地，消减自身，融入环境。

顺形：将建筑形态顺应自然地形，使之与景色协调，也赋予建筑特色。

借势：建筑巧借地势布局，使建筑的形体因地形原本的态势彰显力量，也使建筑成为整体风景的一部分。

靠色：建筑选用与环境较一致的色彩，使之与自然协调。

轻透：让建筑尽量轻巧空透，保持风景视野的连续。

映影：用镜面性材料作为建筑界面，通过映射消隐建筑宏大的体量，将建筑融入风景的同时，将风景的影像引入建筑。

观景：创造不同的观景体验空间。

引景：将景观引入建筑空间，抑或穿透建筑看到不一样的风景。

构景：将建筑空间造型构建成独特的景观。

风景融入
LANDSCAPE INTEGRATION

敬畏自然，保护自然，让建筑融入风景作为景观环境的有机组成部分，呈现自然、质朴、有机的美学是本土设计的策略。其追求的不仅是视觉形象的从属感，更是为人们创造体验自然，观赏风景的空间场所。

用看得见的技术探索看不见的科学
Use visible technology to explore invisible science

日照市科技馆 RIZHAO SCIENCE MUSEUM

设计 Design 2015 · 竣工 Completion 2020

地点：山东日照 · 用地面积：103 786平方米 · 建筑面积：19 633平方米
Location : Rizhao, Shandong · Site Area : 103,786m² · Floor Area : 19,633m²

合作建筑师：关飞、叶水清、胡水菁、高凡、彭彦、伊拉莉亚、冯君
Cooperative Architects : GUAN Fei, YE Shuiqing, HU Shuijing, GAO Fan, PENG Yan, Ilaria Bernardi, FENG Ju

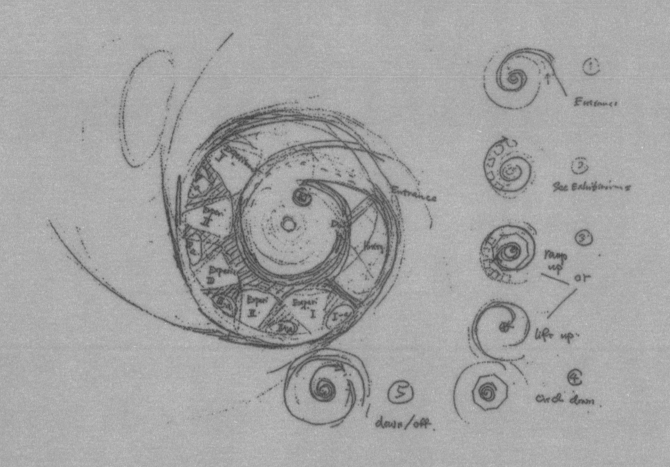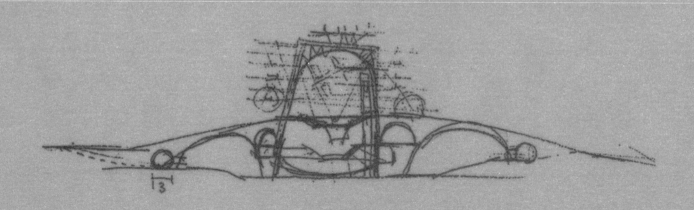

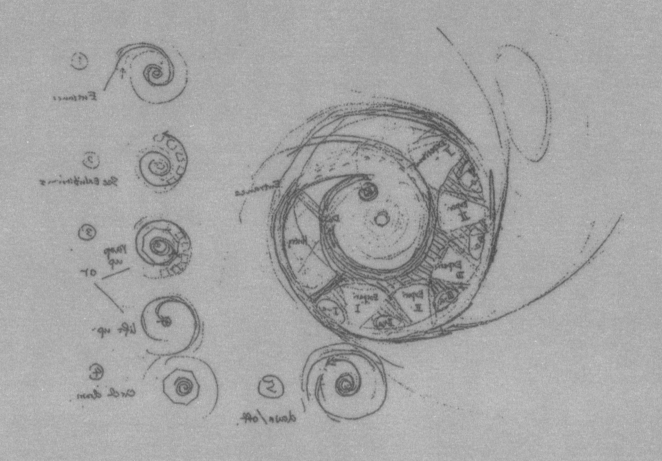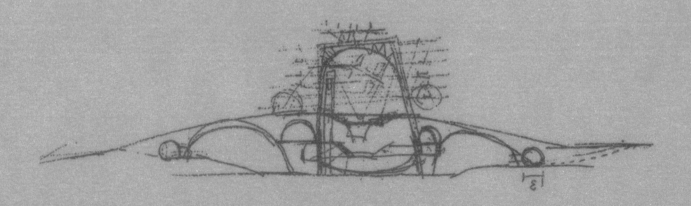

  日照市科技馆展示主题是世界著名物理学家丁肇中（祖籍日照）的科学实验、研究与发现。设计中引入"黑洞"和"星空"的概念，将建筑单体表现为场地景观的一部分：科技馆作为"黑洞"中心，所有场地景观要素及道路围绕中心沿螺旋状放射形布置，并最终导向科技馆的入口空间。通过覆土屋面、融入地景来协调科技馆与相邻的日照会展中心在体量上巨大的反差，形成二者视觉上均衡的相互关系。

  建筑形态在意向上表达"黑洞上的磁谱仪"的概念，分为上下两部分。覆土之下以"黑洞"展厅为中心，以"隧道空间"为媒介串联各实验展厅。这部分在室外表现为微微隆起的覆土地形，让人产生深入地下探索未知的暗示，隐喻探索未知世界是物理学研究的核心和终极价值。覆土部分之上是八边形的轻钢结构体系，包裹着巨幕影院和模型展示区，展示国际空间站和丁肇中团队的重要研究ＡＭＳ实验模型。采用预应力拉索结构，以钢筋混凝土剪力墙支撑屋顶钢桁架结构，悬挂可环视城市景观的坡道和观景平台，展示科学理念的同时也为公众提供了丰富的空间体验场所。

The renowned scientist Professor Samuel Chao Chung Ting, whose ancestral hometown is Rizhao, provided his scientific experiments, research and discovery as the main content of the exhibition of Rizhao Science Museum. The images of "black hole" and "starry sky" are materialized in the entire masterplan of the intervention, landscape and building as a whole, which has been drawn as a circular core element spinning around its center. By sinking much of the building underground and covering the greenery roof, the huge contrast in volume between the museum and the adjacent Rizhao Science and Culture Center Center is coordinated to form a balanced visual relationship between the two. Symbolizing the form of "magnetic spectrometer on a black hole", the building is divided into overground and underground parts. The "black hole" exhibition hall is the center of the underground part, and the "tunnel space" is the medium of the series of experimental exhibition halls. A slightly raised greenery roof gives a hint of exploration of the unknown, which is the ultimate value of physics research. The overground octagonal light steel structure system, enclosing a giant screen cinema and a model exhibition area, exhibits the International Space Station and Ding's team's important research AMS experimental model. Hanged by prestressed cable structure, the ramps and and platforms give visitors fantastic view of the city and seashore, showing scientific ideas and providing an urban space for the public to experience.

日照市科技馆 · 2015—2020 · RIZHAO SCIENCE MUSEUM

嵌入：将建筑的主要展示空间嵌入地下，嵌入滨水公园景观之中，建筑屋顶形成覆土，实现建筑体量的消隐。

隐喻：建筑物露出地面部分是一个由八边形轻质坡道围合的锥台形体，形态上隐喻丁肇中先生的实验仪器AMS02。

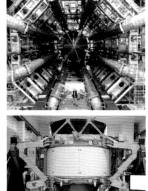 

观景：地上元素由一个围绕建筑核心的螺旋坡道组成，为游客提供360°的周围景观，直到覆盖博物馆主体的山丘上方15m的高度。

标识：科技馆的高度是整个滨水景观空间的制高点，其独特的造型在以绿色为基调的环境里形成了独特的标识。

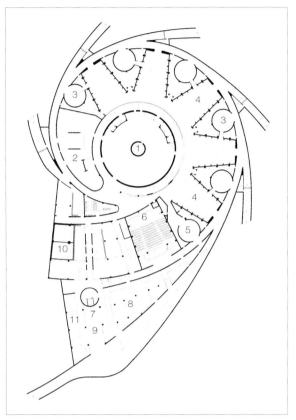
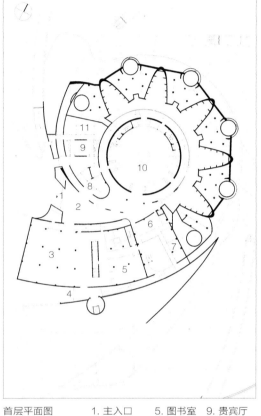

地下层平面图	1. 投影基坑	5. 休息厅	9. 卸货平台
	2. 纪念展厅	6. 报告厅	10. 设备间
	3. 圆形展厅	7. 前厅	11. 辅助办公
	4. 展厅	8. 恒温恒湿库房	

首层平面图	1. 主入口	5. 图书室	9. 贵宾厅
	2. 大堂	6. 会议	10. 序厅
	3. 临时展厅	7. 办公	11. 设备间
	4. 报告厅门	8. 问询处	

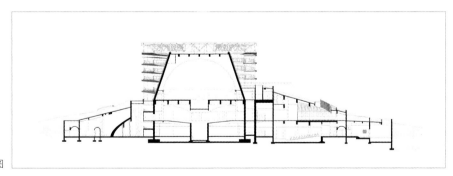

剖面图

LAND-BASED RATIONALISM III 185

日照市科技馆 · 2015—2020 · RIZHAO SCIENCE MUSEUM

日照市科技馆 · 2015—2020 · RIZHAO SCIENCE MUSEUM

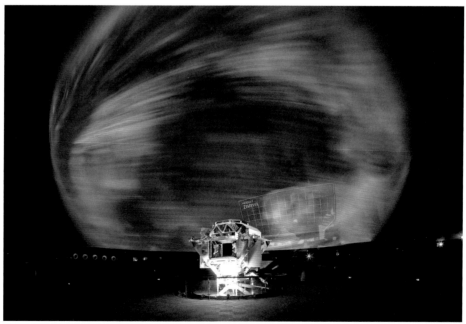

LAND-BASED RATIONALISM III *191*

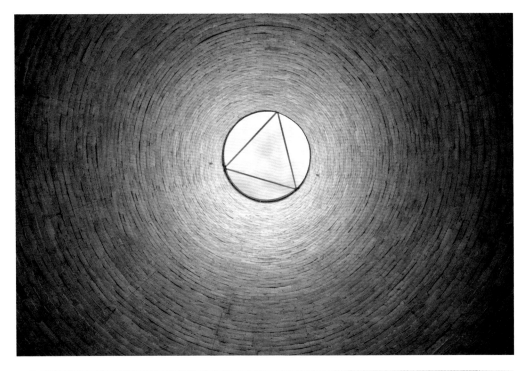

日照市科技馆 · 2015—2020 · RIZHAO SCIENCE MUSEUM

设计随笔

著名科学家丁肇中教授是世界公认的物质和原子粒子物理学领域的杰出学者，因发现J粒子于1976年获得诺贝尔物理学奖。他说："实验是自然科学的基础，没有实验证明，理论是没有意义的。只有实验推翻了理论，才能创造新的理论，而理论不能推翻实验。在过去的400年里，我们对物质结构的基本理解大多来自实验物理学。"

丁肇中祖籍是山东省海岸线上的日照市。从1975年开始，他每年都要回中国访问讲学，为了祖国高能物理的发展，不辞辛劳，努力促进国际物理学界同中国物理学家的合作。2014年，接到日照市建设丁肇中科学馆的委托任务后，崔愷院士立即飞往日内瓦与丁教授会面讨论设计理念和策略。在设计前期丁教授直接参与项目的讨论；在整个设计阶段，他都对科技馆的科学成果、展览材料（设备、模型、实验演示）以及展览的基本内容进行了审核。科技馆展览的内容由丁肇中先生不同阶段的研究成果组成，包括五大实验以及AMS-02球幕影厅。

最初的草图

科技馆的选址位于日照奥林匹克水上公园西侧的公园内，与青岛路相邻。与青岛路西侧的城市用地不同，科技馆更属于公园和海景的一部分，而不是城市。

对于建筑师来说，如何通过空间语言将高深的科学实验同大众的认知联系起来是我们需要面对的核心问题。新的科技馆应该是科学界和大众之间联系的纽带，建筑空间与展陈设计应该将丁肇中先生的科学家精神传递给年轻一代。

崔愷院士最初的草图尝试将动态的星系运动与五大展厅的布局建立联系，所有展示空间都被置入地下，建筑平面布局的核心如同一个黑洞，五大展厅像星球一般环绕着中心，所有的元素都是动态、向心、内卷的，最终都将被吸入到黑洞中。串联展厅的通道甩出长长的轨迹，延伸到整个地面景观中成为地面景观的主要元素。露出地面的球幕影厅是整个科技馆唯一露出地表的构筑物，它象征着一个巨大的AMS带着微小的倾角置于草坡之上，成为整个滨水公园的标志物。

这张草图也让人联想起丁肇中先生模拟宇宙形成的L3实验所在地的巨大圆环，也就是欧洲核子研究中心的地下加速器。这个圆环周长27km，深埋于地下100m。

无论是星系运动还是加速器，设计草图都呈现出一种动态的叙事方式，这意味着建筑要寻求自身的空间语言，与动态的叙事方式同构。

动态的同构

受丁教授实验的启发，设计将景观和建筑作为一个整体，被绘制成围绕其中心旋转的向心环绕型布局，在其两侧排列了入口和出口大厅、五个带卫星室的大型展厅、一个会议厅和服务空间。所有这些空间和循环通道都受到核心"黑洞"的旋转力的作用，因此它们的位置令人想起发动机在主轴上的旋转——"瞬间冻结"——类似于粒子加速器的运动。

动态的平面布局像一个螺旋线圈卷在它的中心，拖着它的尾部元素（与欧洲核子研究中心的环形布局类似）。我们主要的设计理念在于"同构"地表达建筑语言的象征意义：科技馆展览空间像一个巨大的装置邀请游客进行科学探索，流线如同粒子一般沿环形加速运动，进入各个展厅及其子展厅。

游客从±0.000层的入口开始，经过圆形"黑洞"序厅后直接进入地下空间，就像进入黑暗的洞穴一样，主循环路径呈环形展开，围绕核心空间（象征着星体黑洞），穿过螺旋走廊和-6.000层的展厅（对物质的探索），再环绕"黑洞"的环形坡道向上；到达顶部高度后，流线继续将游客带到外部坡道上，然后返回地下，连接到入口前厅。

作为以丁教授的科学研究历程为基础的常设展览由五个拱形展厅组成，分别展示测量电子半径、重光子研究、发现J粒子、揭开胶子之谜、L3实验五大实验设备。它们再现、证明和解释了丁教授完成的五个最重要的实验场景，其中包括各种粒子加速器。每个展厅都根据展示装置和模型的大小进行了设计，所有展厅都被拱形屋顶包围，突出的多个拱门逐渐减小。拱顶的结构混凝土为一次性浇筑，经处理后没有任何表面装饰，并为机电设备提供了空腔。

地上部分由围绕建筑核心的双向螺旋坡道组成，为游客提供360°的连续观景的步道，直到山丘上方15m的高度后可以选择另一个坡道返回。

施工的挑战

如何施工逐渐弯曲的拱顶一直是该项目的挑战，所有展厅及通道的木纹混凝土结构都需要一次浇筑成功，没有失败的余地。项目组通过对BIM模型的全面控制，对每个钢龙骨模板的半径等参数进行了设置和编号；然后，所有弹性胶合板在工厂进行加工，在现场组装和定位。单曲面或双曲面木纹混凝土空间是该建筑地下两层的常见特征：除了五个主拱形大厅外，每个大厅侧面都有一个卫星展厅，形状像顶部削平的圆锥体，人们通过圆形天窗望向天空；入口大厅由一系列同心拱顶覆盖，拱顶由较低且宽的拱门支撑；参观通道的圆形走廊也是拱形的。弯曲的墙壁和拱形屋顶的交叉点，全部由结构和清水混凝土浇筑完成。结合混凝土的预留预埋，室内照明灯位都恰当地隐藏了自身，所营造出来的光环境给游客带来了非常特殊的体验：仿佛是与科研人员一块进入到地下隧道探访隐秘的实验室，增加了空间的神秘感。

中心空间是建筑的核心，也是其主要的旋转中心。它是一个由结构混凝土连续墙形成的锥台形，位于地下圆形展厅上方。在首层入口序厅，中央圆形大厅的地板和天花板向中心略微倾斜，仿佛黑洞的巨力将地面牵引变形。该展厅通过三维投影展示了丁教授对宇宙缘起的研究。序厅的上层是球幕影厅，球幕影厅的中部是一个可以升降的机构，用来展示AMS-02的模型。

地上锥体的外表面覆有镜面反射波纹铝板，直径18m，高度18m。整个地上部分的重量由混凝土锥体支撑，其周围的坡道钢结构仅由拉杆锚定在地面。锥体本身构成了支撑屋顶和螺旋坡道的主体结构，其结构系统可以通过三个不同的部件分开来理解：混凝土锥体的顶部是一个巨大的钢桁架屋顶，上部覆盖ETFE膜材料采光屋面；坡道通过一系列绷紧的钢拉杆由屋顶的边梁吊挂着，以双螺旋的方式上升。该结构系统允许将拉杆的截面减小到最小，以代替较粗的钢柱来支撑重型坡道的平台板。

混凝土圆锥体的曲面和圆形坡道的外沿是八边形的檐口，它与屋顶八边形轮廓相对应。八边形钢结构的造型来源是安装在欧洲核子研究中心实验室的大型强子L3对撞机，这是丁教授构思和建造的最重要的设备之一。

从一定距离看，科技馆给人的第一印象是一个轻盈、通透的物体，由镜面的金属材料和钢索组成，微微倾斜的八角形屋顶、坚实的镜面反光锥以及螺旋坡道。相对于建筑物，它更像一个巨大的实验装置，仿佛突然降落在绿色的地面上，引起了不知情路人的注意和好奇。在夜晚，这个"天外来客"所发出的奇异光线更是日照市滨海区域的一个神奇的景观。

后记

根据丁肇中先生的建议，科技馆的名称从丁肇中科技馆改为日照市科技馆。这标志展览的范围和内容的扩大，也代表科技馆的展览目标不仅是面向研究学者，更是面向广大学生和普通访客的。就像前文所说的：日照科技馆能否成为衔接实验物理学与广大民众的纽带，值得期待。注8

绿色的浪
Green waves

日照科技文化中心 · RIZHAO SCIENCE AND CULTURE CENTER
设计 Design 2017~2018 · 竣工 Completion 2023

地点：山东日照 · 用地面积：449 451平方米 · 建筑面积：440 437平方米
Location : Rizhao, Shandong · Site Area : 449,451m² · Floor Area : 440,437m²

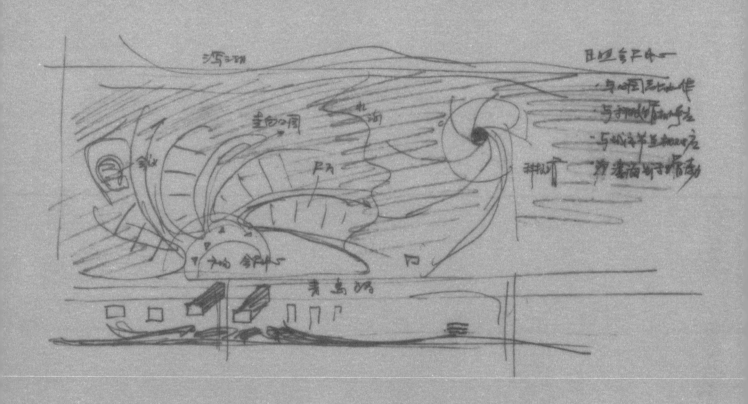

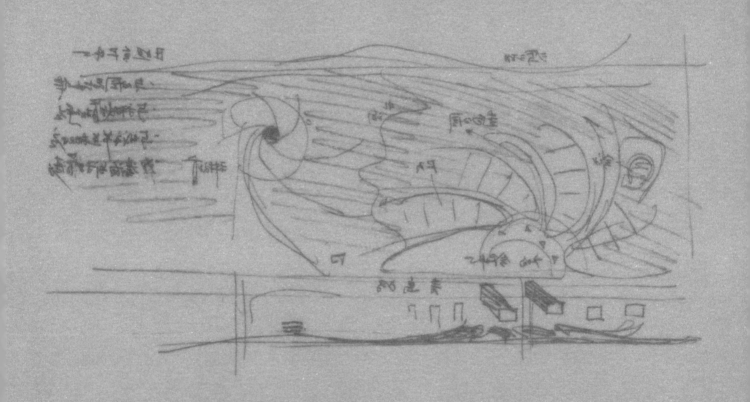

项目位于日照市奥林匹克水上公园内，建筑布置和空间形态上延续了公园绿地的景观特征，突出与相邻的日照市科技馆的整体性以及核心广场的向心性，利用流线型交通体系串联广场和建筑。建筑群沿城市界面以富有韵律的形态缓缓展开，以广场为核心展开各功能布置，包括展厅、大剧院、游客中心、商业、酒店等，体现城市建筑的体量感，利用玻璃和反射材质的交互映衬，在不同时刻的光线照射下呈现出富有变化的光影效果和建筑形象。面海一侧，主要功能被置于绿坡覆土之下，屋顶的运动健身、观景平台、开放的公园化景观将城市功能融入滨海风光带之整体中，排洪渠功能与景观功能融为一体，凸显"生于自然，融于自然"的设计理念，让人感受原始地貌的亲近和友好。

Located in Rizhao Olympic Water Park, Rizhao Science and Culture Center has adapted the landscape features of the park and has highlighted its connection with Rizhao Science Museum, with a streamlined transportation system connecting squares and buildings. An exhibition hall, a grand theater, a tourist center, commercial spaces and a hotel are arranged around the square, creating diverse light and shadow effect with various materials. On its side facing the sea, main functional spaces are covered under green slopes, upon which spaces for fitness and scenery-viewing are merged with the costal landscape. Flood drainage channels are integrated with the landscape to highlight the concept of harmony between the buildings and nature.

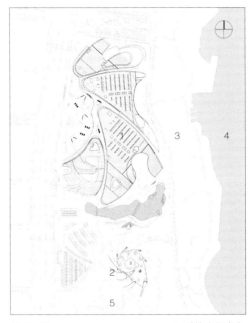

总平面图

1. 大剧院
2. 多功能展厅
3. 展厅
4. 未来馆
5. 酒店
6. 商业
7. 下沉庭院

首层平面图

1. 科技文化中心
2. 科技馆
3. 水上公园
4. 黄海
5. 日照植物园

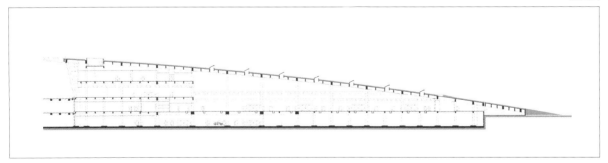

剖面图

LAND-BASED RATIONALISM III *199*

日照科技文化中心 · 2017—2023 · RIZHAO SCIENCE AND CULTURE CENTER

日照科技文化中心 · 2017—2023 · RIZHAO SCIENCE AND CULTURE CENTER

日照科技文化中心 · 2017—2023 · RIZHAO SCIENCE AND CULTURE CENTER

黄河的痕迹
Trace of the Yellow River

兰州市城市规划展览馆　LANZHOU PLANNING EXHIBITION HALL
设计 Design 2012 ・ 竣工 Completion 2016

地点：甘肃兰州 ・ 用地面积：11 428平方米 ・ 建筑面积：16 270平方米
Location : Lanzhou, Gansu　Site Area : 11,428m^2　Floor Area : 16,270m^2

兰州市城市规划展览馆位于黄河北岸，用地呈不规则状沿黄河展开。建筑取"黄河石"为设计意向，经过切削的体块犹如一块被石头包裹的黄河璞玉，经过河水经年累月的冲刷，呈现历史和文化的沉淀。立面采用粗犷的现浇清水混凝土，水平向的凹槽自下而上间隔渐宽，南立面还嵌入多处横向的玻璃嵌缝，既表达河水冲刷的感受，也化解了大面积混凝土材料的单调感。这些玻璃嵌缝也成为从建筑内部欣赏黄河美景的窗口。近人尺度的凹缝内还印有黄河卵石肌理，使建筑的地方意味更加浓厚。河堤部分延续整体的折面形态，与立面的水平向条纹肌理一致。东端的十二面体的宝石形景观挑台，为人们提供了眺望黄河的位置。中部清水混凝土折面向河滩撕开，形成平缓的踏步，为沿河步行的市民提供下至河滩的便利路径。内部空间采用回字形的展陈流线围绕城市总规模型展开，外部的清水混凝土墙面也延伸至室内。

Located on the north bank of the Yellow River, the building resembles a stone washed by river for months and years. By means of cutting, it looks like an unprocessed jade covered by stone that indicates the rich culture of the historical city Lanzhou. All exterior walls are cast-in-place concrete to produce an impressive texture. The horizontal grooves arrayed with varied spacing, especially several horizontal glass strips on the south façade, give its appearance more richness. Those glass strips are also viewing windows of the river. Some grooves with pebbles inside have added to the local characteristics of the building. Folded surfaces are also seen on the river walls, with texture identical to that of the façade. A diamond-shaped landscape platform offers a view of the Yellow River. The folded surfaces have been "split" to accommodate steps down to the bay. The interior space uses a zigzag circulation line to unfold around the general scale model of the city, and the external concrete wall extends into the interior.

兰州市城市规划展览馆 · 2012—2016 · LANZHOU PLANNING EXHIBITION HALL

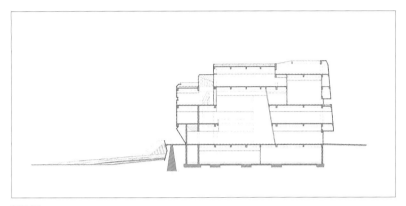

剖面图

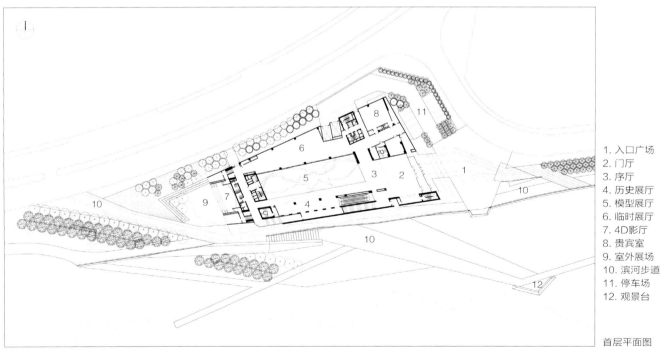

1. 入口广场
2. 门厅
3. 序厅
4. 历史展厅
5. 模型展厅
6. 临时展厅
7. 4D影厅
8. 贵宾室
9. 室外展场
10. 滨河步道
11. 停车场
12. 观景台

首层平面图

兰州市城市规划展览馆 · 2012—2016 · LANZHOU PLANNING EXHIBITION HALL

LAND-BASED RATIONALISM III 213

让山留在镇上
Manmade hills in the town

崇礼中心 · CHONGLI CENTER
设计 Design 2019 · 竣工 Completion 2022

地点：河北张家口 · 建筑面积：24 035平方米
Location : Zhangjiakou, Hebei · Floor Area : 24,035m^2

合作建筑师：康凯、王庆国、张一楠、李俐
Cooperative Architects : KANG Kai, WANG Qingguo, ZHANG Yinan, LI Li

策 略：覆土、嵌入、轻透、观景

摄影：张广源、李季
Photographer: ZHANG Guangyuan, LI Ji

借冬奥会契机打造的崇礼中心包含冬奥和冰雪主题博物馆、崇礼区图书馆和一个综合文化广场，既是向世界展示中国冰雪运动发展的重要窗口，也成为这座生态宜居城市的活力之源。

基地位于城市与连绵起伏的群山之间，宽200m、长400m的开阔坡地东高西低，有近30m的高差。建筑舒展的屋顶曲线配合周边的大尺度环境，强化出"山"的轮廓。屋面中嵌入的三座通体晶莹的玻璃厅，让光线射入建筑深处，宛若三座巨大的冰晶从"山体"上破土而出。

博物馆和图书馆顺"山"势设于面向广场一侧，博物馆在下，图书馆在上。主要功能平面由三角形单元首尾排列组成，三角形的几何空间逻辑从首层延续至上层并最终穿出屋面。博物馆面向城市一侧的清水混凝土肋柱形成抵抗侧向推力的整体排架，清晰的竖向线条在外观上与山脚下的白桦林相映成趣。逐渐隆起的建筑体量与向中心汇聚的层层白桦林带围合出城市广场；纵横交织的慢行步道将人流从周围引入，并从广场延伸到建筑上部，其间形成高低错落的景观台地。五个巨大的白色拱门跨越廊道，形成具有仪式感的广场入口。

Consisting of a winter sports-themed museum, Chongli Library and a comprehensive cultural square, Chongli Center, built on the occasion of Winter Olympics 2022, is both a window to the world and a source of vigor for the livable city.

Located between the city and undulating mountains, the 200 x 400 meter site descends from the east to the west with an altitude difference of nearly 30 meters. The roof has a stretched outline resembling the mountain, with three crystal clear halls inserted to introduce light into the building, presenting an image of three giant ice volumes breaking out of the "mountain".

A museum and a library face the square, with the museum located at a lower position than the library. Main functional spaces are consisted of triangular units, which are arranged throughout the whole building. An array of fair-faced concrete ribbed columns are arranged at one side of the museum to resist lateral thrust. A square is formed with the rising volume of the building and birch woods, which can be accessed from pedestrian lanes. Five huge white arches stretch over a path, forming a well-marked entrance for the square.

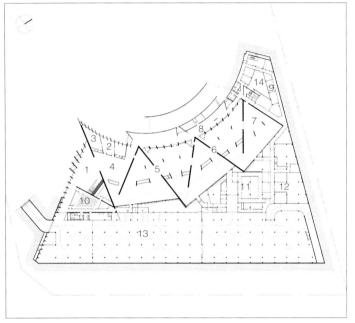

博物馆层平面
1. 公共大厅
2. 贵宾门厅
3. 贵宾室
4. 序厅
5. 冰雪展厅
6. 奥运展厅
7. 互动展厅
8. 环廊
9. 办公
10. 多功能厅
11. 展品库房
12. 设备机房
13. 半地下车库
14. 庭院

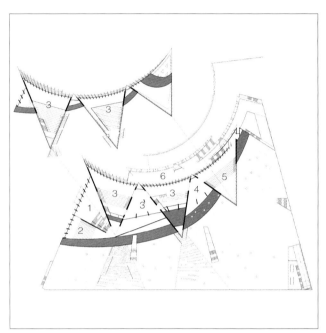

图书馆层平面
1. 公共大厅
2. 书店
3. 开架阅览区
4. 少儿阅览区
5. 多功能阅览区
6. 室外平台

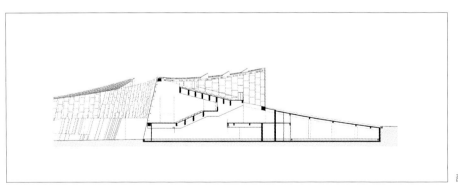

剖面图

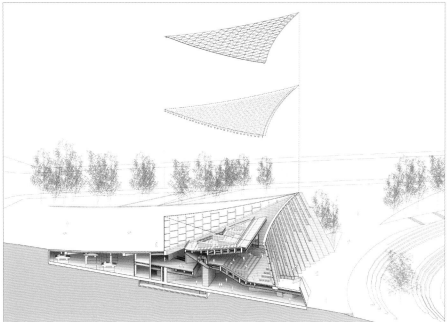

剖透视图

崇礼中心 · 2019—2022 · CHONGLI CENTER

崇礼中心 · 2019—2022 · CHONGLI CENTER

设计随笔

2015年7月，随着北京成功申办2022年冬奥会，中国继2008年北京夏季奥运会之后，再次成为世界瞩目的焦点。作为冬奥会雪上竞技项目主要赛区的崇礼，也迎来了城市建设和冰雪产业发展的巨大机遇。七年之间，生态修复，旧建翻新，使原本平淡无奇的城市风貌焕然一新。加上冬奥会的知名度和雪场的聚集效应，越来越多的人慕名而至，这个曾经名不见经传的小城摇身变成世人向往的"冰雪天堂"。借冬奥会契机打造的崇礼中心是"两馆一场"的综合文化项目，包含冬奥和冰雪主题博物馆、崇礼区图书馆和一个大型综合文化广场。崇礼中心作为奥运会留给崇礼的重要文化遗产，既要成为向世界展示中国冰雪运动蓬勃发展的重要窗口，也要兼具对地域特色的发掘和对日新月异生活方式的探索，并最终在城市日常性语境下成为生态宜居的城市活力之源。

城里的"山"

项目位于连绵起伏的群山脚下，风景资源得天独厚。用地是一处宽200m、长400m的开阔坡地，地势东高西低，落差近30m。面对这片背山面城、有着强烈山地属性的场地，创造一处融于自然的大地景观似乎是题中应有之意。于是，在场地里创造一座"山"便成为方案的出发点。

为了不让这座"山"成为单纯模仿自然地形的覆土建筑，而呈现出建筑与场地相互兼容、相互拟合之后"顺其自然"的状态，我们首先从周边的环境关系着手对场地进行布局，在不同方向上采用不同的态势和边界处理来塑造形体。

用地东侧靠山的东环路是进入城区的快速路，平时鲜有人至，我们将建筑紧邻道路布置，使形体成为与用地边界连成一体的坡地造型；屋顶则作为自然山体的视觉延伸，从地面缓缓抬升延展至西南角高高扬起。南侧形成垂直于地面的断面，逐渐旋转打开的列柱形成礼仪性的入口。西侧的裕兴路靠近城市，是南北贯通的主要干道，也是人流的主要来向，于是我们将文化广场放在建筑与道路之间。屋顶舒展的曲线配合周边的大尺度环境，强化出"山"的轮廓，形体微微内凹，形成对广场的边界围合感。为了强化面向城市的开放性，我们用宽窄、疏密和偏转角度不同的肋柱在立面上排列出富有节奏感的韵律，随着观看视角的不同形成虚实过渡的变化，也随之化解了形体的厚重感。为了强化从高出场地10m的张承高速上远眺建筑的效果，我们在面山一侧的屋面嵌入了三个通透的玻璃厅，既能让充足的光线射入建筑深处，又能营造出富于震撼效果的第五立面，从空中俯瞰宛若三座巨大的冰晶从"山体"上破土而出。

当起伏的山地造型成为一个由外而内的环境策略，接下来要解决的就是让内部空间与其功能契合。我们顺"山"势将博物馆和图书馆功能放在面向广场的一侧，博物馆在下，图书馆在上。整个建筑不做地下室，车库、库房和机房等附属用房则利用场地高差嵌入临东环路一侧，形成可以通风和采光的半地下空间。主要功能平面由三角形单元首尾排列组成，三角形的几何空间逻辑从首层博物馆一直延续至上层图书馆并最终穿出屋面。上下一致的空间形态让结构墙体既实现了竖直方向的支撑，又在平面内与流线组织、空间表达关联在一起。在博物馆中，连续转折的结构墙体将展览空间依次切分成不同主题的展厅，形成了转折迂回的观展路径。两馆合一的入口大厅是夹在两片通高结构墙体之间的楔形空间，为了让进入大厅的人有走进"山"洞的感觉，将墙和顶以相同的界面来处理，整个空间浑然一体，呈现出极具视觉冲击力的纵深感。考虑到屋顶是逐渐抬起的缓坡，在图书馆中，如果竖向平层分区，为满足屋面最低位置的空间净高，各层楼板不可避免要大幅度内收，将大大降低空间使用效率。我们将楼板处理成不同角度的斜坡，上下层之间互为地面与屋顶，一方面形成了丰富奇特的空间变化，同时也为读者提供了远近高低各不同的观景视野，朝南俯瞰城市，朝东眺观山景。

从大厅台阶往上走，即进入图书馆。现代图书馆对于未来城市生活而言不仅是一个藏书和阅读的场所，更是一个让人驻足忘返、交流互动、触发灵感和惊喜的空间。作为"山"里的图书馆，我们希望在这里看书的人都能感受到山的气息。层叠错落的坡地宛如一片片抽象的山地景观覆盖并渗透到整个图书馆空间，温暖的天然橡木条板包裹地面和天花，营造出静谧的氛围和舒适感。为了让阅读变得更有打卡的体验，我们在不影响通行的前提下，将台阶处理成可以落坐的台地；台地上分布着藏书、阅读和社交的功能区，通过半人高的书架将彼此的视线隔开。面向城市一侧，宽大的清水混凝土肋柱逐渐扭转阵列展开，光和景渗入馆内，同时有效屏蔽了西向的强光。拾级而上来到面山的台地，两侧纤细的白色钢柱撑起反曲屋面，通透的玻璃界面让远处的景色渗透室内，也将内部丰富的活动场景悉数展现给城市。为了让突出屋面的反曲穹盖无论从空间内部向外看还是通过玻璃界面向室内都足够轻巧纤薄，我们将空调系统结合落地书架设计。排烟方式也采用顶部自然排烟，屋面仅保留必要的做法并将三角形网格结构完全暴露在空间中。散布在屋顶通过消防联动开启的自然排烟口同时兼顾夏季通风功能，可有效地排出聚集的热气。

融入生活的生态地景

在城市公共空间营造中，我们希望将这处依山傍城的场地打造成城与山相互融合的大地景观，既有大山的野趣，也不失城市的理性，并以开放的姿态面向未来多元化的城市生活。

位于用地中心的城市广场，围合于建筑从大地隆起的庞然体魄与逐渐向中心汇聚的层层白桦林带之间。林下的山花和草丛勾勒出纵横交织的慢行步道，将人流从四面八方引入广场。原本设计了一些磨光料石点缀在广场中，随着光线和视角的变化，地面上便会形成斑斓闪烁的肌理，夜晚在灯光的映射下，效果更为明显。但考虑到磨光的石面在雨雪天会异常湿滑，最后只好放弃，继而改为白色和深灰色毛面花岗石石板的搭配混铺，虽不及原来变幻的视觉效果，但也勾画出灵动的笔触，进而形成了雪涡的趣意。考虑到每逢大型活动，广场中间就会搭起舞台，我们便将错落有致的坐凳嵌入缓坡，形成若干组环抱广场的看台，厚重的花岗石条石搭配菠萝木饰面，美观又耐用。看台构成的漫步通道从广场一直延伸到建筑，进而形成建筑底部高低错落的景观台地，人们可以沿着曲折的台阶和坡道登上高处的平台驻足远眺。

为了让东侧的大屋面不单单只是从远处观赏的雕塑景观，而能更贴近生活，我们将其坡度较缓的区域全部对外开放，人们可以随意走上屋面，或席地而坐、远眺风景，或悠闲漫步，近距离感受屋面作为大地景观的壮阔尺度。每逢雪季，屋面又会成为巨大的雪坡，引得孩子们爬上滑下、戏雪玩耍。建筑背坡原本就呈现出柔和的自然形态，试想如果再将屋顶布满绿植，那么看上去就无异于远处的绿色山体，不免乏味单调。于是我们转而采用不同深浅和肌理的石板将其覆盖，

铺装的做法也随位置不同而变化。随着屋面缓缓抬升，由接地处深灰色自然面石板散铺，逐渐过渡到顶部白色麻面石板密拼。天长日久，石缝中便会自然长出绿草，在屋面上形成一片退晕的绿色，远远望去，既感与远山相融，又萌发雪峰的联想。

夜晚，我们刻意将环境光压到最低限度，不与星空争光，营造世外桃源般恬静的氛围。建筑的光也是隐匿的，安静自然地"长"在建筑上。面向广场，嵌入地面的窄角度光源勾勒出混凝土肋柱挺拔的线条和精致的边角。灯光亦可变化，配合夜晚广场的活动，形成如音符跳跃般的动态节奏。面山一侧的三个玻璃厅内则是灯火通明，强化光自内向外溢出的效果，使之成为坡地上熠熠生辉的"冰晶"。建筑入口一侧耸立着高大的清水墙，为了中和单一的表面肌理，我们将一组或高耸挺拔、或舒缓延展的造型油松点缀在旁边。夜晚，随着灯光亮起，斑驳的树影洒向清水墙面，晕染出朦胧的诗意图底。

回顾整个项目的设计与建造历程，虽然因其启动较晚而未被列入奥运项目名单，但这也恰恰摆脱了建筑作为彰显国家形象的宿命，让设计返璞归真，呈现出其应有的状态——建筑生于自然，又融入自然——使存在的结构隐于无形，而让隐藏的逻辑与秩序由内而外悄然显现。每当风和日丽，这座小城里的"山"便会沐浴在明媚的阳光之中，在山色的四季轮转之间，幻化出千变万化的表情，安静而又充满活力。"山"与城互为孕育，和谐共生，这也许才是它最打动人的地方吧。注9

关爱：在满足基本任务功能的前提下要关注各类使用者的需求，让建筑的服务功能和细节设计更加方便和友善。

共享：调动建筑中主体功能之外的辅助空间，让它成为人们遇见、交流和引发非正式功能的场所。

共建：让空间和建筑跨越权属边界，使城市公共联系更连续，公共空间能延伸，这需要设计的推动和投资的融合以及管理的创新。

分享：让原本内部使用的功能和空间向市民或邻里开放，这有赖于这些空间布局有利于开放。

混搭：将多种功能和社会关系合理混合搭配，会有助于提升城市活力，减少交通能耗，也能促进社会公平。

开放：努力用设计的方式让业主单位自用的空间具有公共的属性，而这个公共性应该是有质量的公共场所，让"公私"双赢。

场景：将内外空间场景化，形成主题性体验场所。

交往：让功能空间和流线的组织有利于人与人的遇见。

消费：将消费功能加入其他功能空间，促进消费，提升服务质量。

生活引领

LIFESTYLE ADVOCATION

关注社会生活，跟进需求变化，创造人们喜爱的场所，促进公共空间的共享、公平和善意，是本土设计以人为本的基本策略。

纪念是一种传承
Remembrance is a kind of inheritance

天津大学新校区主楼 · MAIN BUILDING OF TIANJIN UNIVERSITY NEW CAMPUS
设计 Design 2012 · 竣工 Completion 2015

地点：天津津南 · 用地面积：139 472平方米 · 建筑面积：85 928平方米
Location : Jinnan, Tianjin · Site Area : 139,472m^2 · Floor Area : 85,928m^2

合作建筑师：任祖华、梁丰、曹洋、叶水清、彭彦、冯君
Cooperative Architects : REN Zuhua, LIANG Feng, CAO Yang, YE Shuiqing, PENG Yan, FENG Jun

策 略：共享、开放、共建、分享、场景、关爱

摄影：张广源、傅晓铭、李季
Photographer: ZHANG Guangyuan, FU Xiaoming, LI Ji

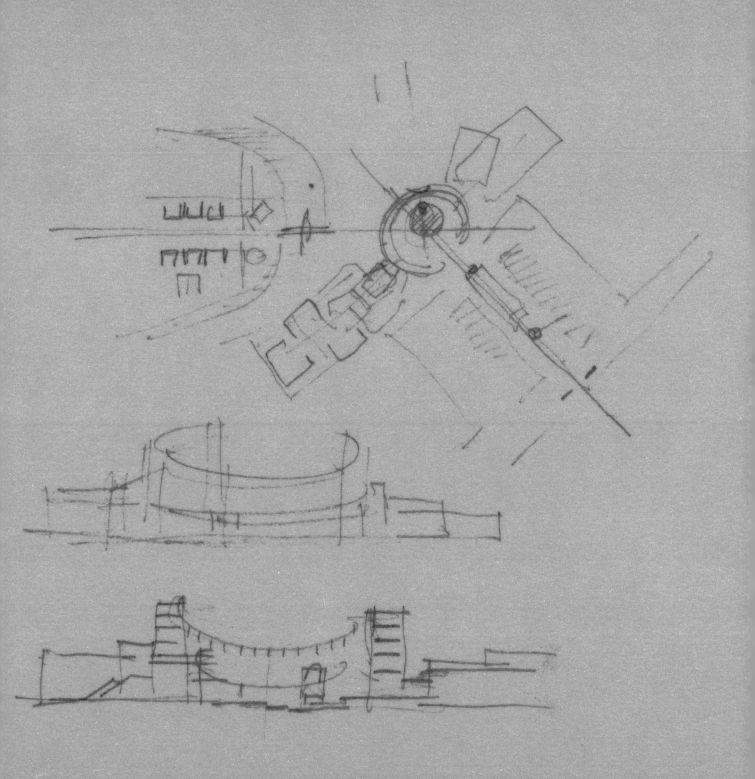

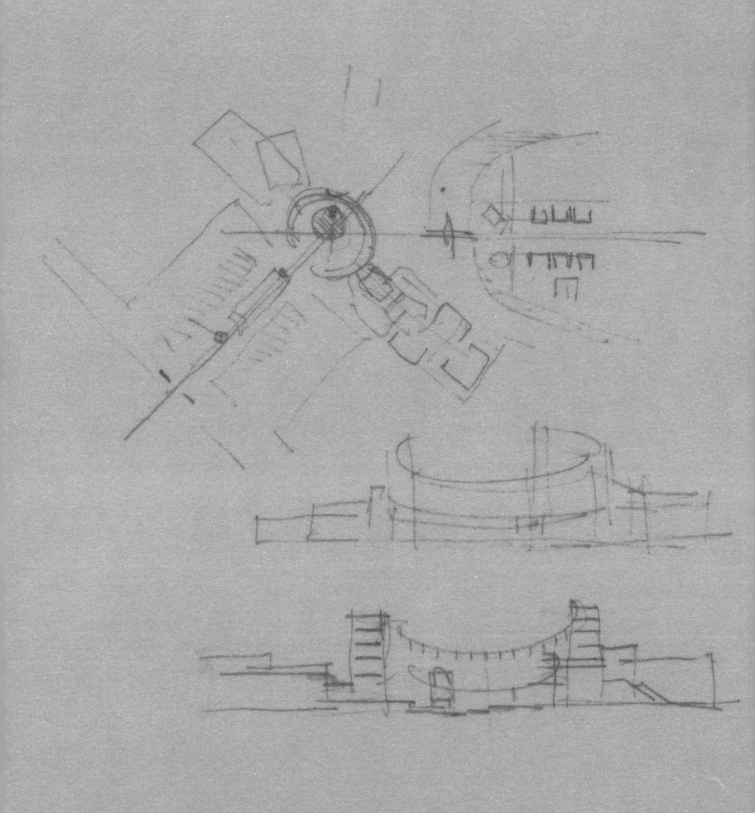

天大新校区主楼是从东侧主校门进入校区的第一组建筑。东西向的校内主轴线与东南向的校前区轴线交汇于此，交点处置入由环形建筑围合的中心广场，使两条轴线完成自然的转换和衔接。学生的聚会交流空间——中心广场成为整组建筑的核心，在东西两侧打开，将东西向的校园步行体系串为一体；环形建筑作为广场的界面和背景，将传统校园主楼从以"楼"为主转变为一种以"学生活动"为主的新校园空间模式。北洋会堂和理学院布置在环形主楼之内，南北两翼分别设置由文法学院、马克思主义学院、职教学院及第四教学楼组成的文学组团和以实验室为主的材料学院，各种复合化的教学功能通过大大小小的庭院和层层的室外平台连为一体，为不同院系学生的交流和共享提供了可能。

Located on the eastern end of the main axis of the campus, the main building is the first structure to be seen from the entrance of the campus. Since the building is situated at an intersection of three axes of the campus, a circular square, with its center at the point of intersection is planned. The main axis goes through the buildings to form diverse open spaces, and the main building is divided into three parts. As the comprehensive laboratory building serves as an important node on the main axis, the two lobbies on both sides of the axis are located symmetrically. Carrying forward the existing features of the university, the façades are dominated by shale bricks. The positions for hollowed-out parts and window openings are arranged in a unified style with slight variations.

天津大学新校区主楼·2012—2015·MAIN BUILDING OF TIANJIN UNIVERSITY NEW CAMPUS

共享：无论是师生还是访客，无论是集会还是庆典，无论是纪念还是欢聚，都可以在这里发生与共享。

场景：广场为场，北洋亭为景，入场观景既是一种视觉的享受，更唤起内心的情感，这就是场景的精神价值所在。

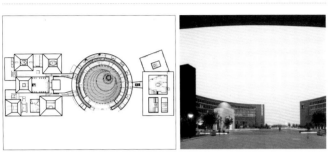
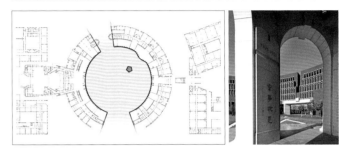

共建：多种教学功能集群混合布置，相互连接，共享公用，提高了空间的使用效率，促进了不同院系之间的交流合作。

分享：室内舞台可向室外剧场开放，室内的演出与室外的同学分享，建筑为围观提供了条件，是对分享需求的响应。

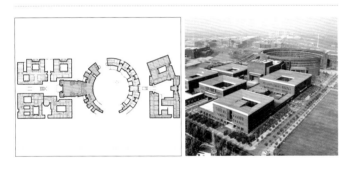
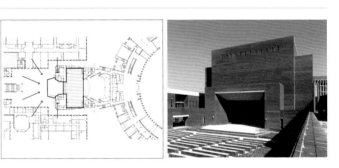

开放：主楼是一个容器，空间是它的内核，而空间并不封闭，一端迎向校门，一端通往校园。开放的建筑让人不是敬畏绕行，而是被迎来送往。

关爱：围合的空间、宜人的尺度、环绕的绿树、树下的长凳、脚下细致的铺装、温馨有趣的光影、庄重而灵动的纪念亭，都会让人停下脚步，体会到百年学府的温度。

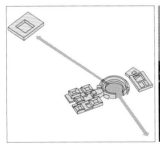

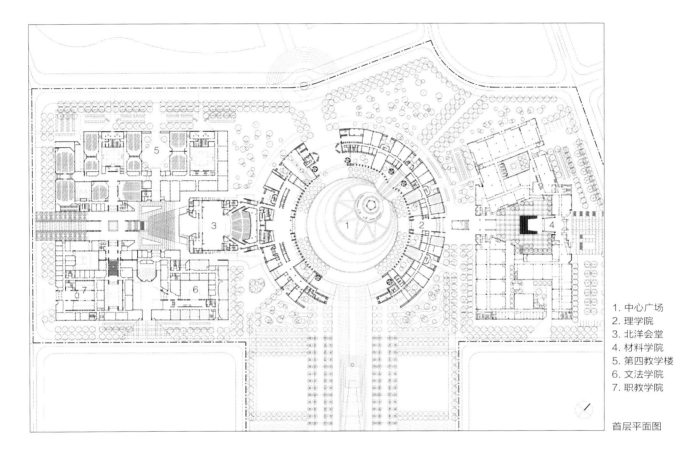

1. 中心广场
2. 理学院
3. 北洋会堂
4. 材料学院
5. 第四教学楼
6. 文法学院
7. 职教学院

首层平面图

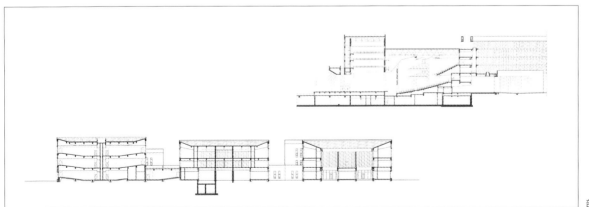

剖面图

天津大学新校区主楼·2012—2015 · MAIN BUILDING OF TIANJIN UNIVERSITY NEW CAMPUS

天津大学新校区主楼·2012—2015·MAIN BUILDING OF TIANJIN UNIVERSITY NEW CAMPUS

设计随笔

为母校做设计是我的荣幸。2015年，在全国高校新校区建设的热潮差不多过去之后，天大也启动了新校区建设的工作。感谢母校对自己学子的信任，让我和周恺担任总建筑师、彭一刚先生担任总顾问。校园的规划经过集思广益，最终选用天津华汇规划院黄文亮先生的方案，并组织众多优秀建筑学专业的校友集群设计，形成"相互交流、统一把关"的集体创作新模式。

天大主楼是校园的前区，位置最重要，设计难度也最大。其中最难的是校门的东南向与校内主轴的正东向之间的偏转角，两条轴线在主楼处要完成转换和自然的衔接。经过多轮多方案的比较，最终选择以圆环状主楼围合中心广场的方案，彭先生亲自设计的北洋纪念亭落位其中，如画龙点睛。广场向东南开口，与校门广场相接，如伸开的双臂，迎接天大的学子和访客；广场正西向开门洞和凹缝，如同标尺提示东西向的主轴线，也许还可以理解为北洋大学堂是东西方文明交流最早开启的大门，有几分纪念性。圆形广场不仅巧妙地解决了轴线的转折问题，也成为北洋会堂和周边教学楼的共享中庭，诠释了以学生为中心的理念。环状大楼高七层，立面分为三段构图，每段的进深、节奏、尺度和开窗方式都有变化，相互叠套，强化环形动感的同时也使其更加稳重并富有层次感。外墙材质选用深咖色页岩实心砖砌筑，强调百年校园的永久性和高品质，也更将白色的纪念亭衬托出来，形成强烈的反差，营造出艺术的氛围。广场铺装多用粗犷的马蹄石，有历史感，部分表面抛光，在日夜的光映下宛若群星闪烁。大树环绕周边、草坪托起亭台、门洞上方是"1895"天大元年标识、让校友合影留念的石刻校徽、通向校门的水池、旗杆广场以及彭先生设计的校门及记时碑刻，每一处都有文化的象征意味，层层铺垫，共同构成了别有生面的前导序列。

环形主楼如同一个巨型的轴承，不仅自身可调方向，还有效地将两翼学院楼连为一体。南侧是文理学院、外国语学院和马克思主义学院，格局采用经典的套院组合，不同的院落有不同的特点，穿越其中有种我们所熟悉的那种传统的秩序感，同时它们也是诸多讲堂、教室的共享庭院。北侧是材料工程学院，大小不同的实验室穿插组合，也构成了大小不同的院落，只不过更加强调效率和理性。

除了天大主楼组团，我们的工作也向学校主轴方向延伸，跨越环岛河的"船桥"，过桥后是方圆两座建筑——公共实验楼和计算机楼，尤其是后来为了纪念母校"天大之星"的计划，设计了一枚星形雕塑，将其置于河面上，夜晚投光灯下，成为耀眼的景观，宛如天上的明星落入校园般闪闪发光。在中轴线端部，我还为77级、78级校友设计了纪念墙，一块块刻有各专业名称和二维码的清水混凝土砌块叠成折形的墙，一个方向指向老校区，一个方向顺着新校区的轴向，两个校区在此交汇于一处，代表校友们对母校的牵挂，而砌成的墙如同大厦的奠基石，象征我们"文革"后第一届大学生建设现代化国家的决心。

三年疫情，让我无法进入我们亲手绘就的校园，但心中常常挂念。彭先生还招呼我们几个弟子去家中，讨论他亲手绘制的张太雷纪念园，几近九十高寿的老人画功不减，密集排列的铅笔线帅直有力，为新校区再添上精彩的一笔，令我等弟子们钦佩不已！这让我想起另一位恩师聂兰生先生，她在告别人生之际拉着李兴钢的手艰难地道出心声：我是爱天大的！是的，我们天大的建筑学人一代代为母校建设做的工作就是我们献给母校的拳拳爱心。从徐中先生的天大九楼、图书馆，到彭先生设计的建筑馆和王学仲艺术研究所，还有学长布正伟先生设计的七里台校门、周恺大师设计的冯骥才文学馆，还有我在天大设计院的多位同学校友们在校园中留下的诸多作品都成为广大天大师生和校友们喜爱和牵挂的"乡愁"。相信我们在新校区的工作也能够延续母校的文脉，成为新时代天大人喜爱的新经典！

开放的建筑 生长的校园
Open buildings in growing campus

北京邮电大学沙河校区　SHAHE CAMPUS OF BEIJING UNIVERSITY OF POSTS AND TELECOMMUNICATIONS
设计 Design 2012　竣工 Completion 2014

地点：北京昌平　用地面积：14 950平方米　建筑面积：35 500平方米
Location : Changping, Beijing　Site Area : 14,950m²　Floor Area : 35,500m²

　　校园规划调整以尊重高教园区的街区尺度和路网体系为出发点，将"一横一纵"的市政道路作为校园的基本骨架和形象轴线，分为"东南西北"四个区域，分别作为宿舍、教学、运动和景观之用，各类公共综合建筑则居于中部。单体建筑以合院式布局为主，不同尺度的院落空间连缀起来形成丰富的校园空间，结合场地中的保留树木形成绿荫遮蔽的街道和庭院空间，高大的白杨树和旱柳延续了场地原有的场所记忆。综合楼采用U字形围合式布局，开口设在西侧，中间容纳了报告厅和两个庭院，形成了内聚的气势和东西向轴线。体量由东向西跌落，屋顶绿化平台间以大楼梯相连。

　　在图书馆设计中，保留的东西向林道和高大的杨树林将场地划分为四个象限，底部以方形体量占据树林之外的三个象限，容纳报告厅等功能，以红砖和深窗洞产生厚重感，窗洞间的壁龛提供了小尺度的阅览空间；上部插入以共享大厅和藏阅空间为主的圆形体量，采用玻璃和铝格栅体现轻盈感，透射出阅读和交流的场景。

The planning of the campus has been carried out with respect for the scale and road system of the region. The campus is divided into four parts, which are used for living, teaching, sports and landscape respectively. Public buildings are located in the middle of the site, and courtyards with varied scales are planned. The comprehensive teaching building descends from the east to the west, forming green terraces connected by grand stairs, which can serve as both a communicating place and the access to classrooms. Classrooms line along the interior corridors, where smoke exhausting wells and vertical windows of glass tiles are located in double walls.

The library located at the northeastern corner of the intersection of two main axes, made great efforts in preserving the poplar woods and a shady path on the site. The lower part of the library takes on a square form and consists of lecture halls, offices and a cafeteria, while the circular-shaped upper part is composed of a lounge and book storage spaces.

北京邮电大学沙河校区 · 2012—2014 · SHAHE CAMPUS OF BEIJING UNIVERSITY OF POSTS AND TELECOMMUNICATIONS

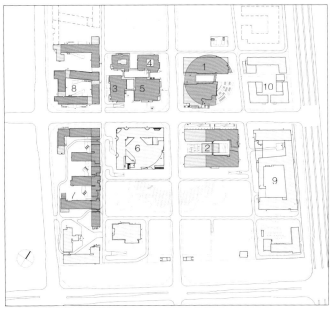

1. 图书馆
2. 教学综合楼
3. 食堂
4. 联合办公楼
5. 活动中心
6. 公共教学楼
7. 宿舍
8. 公寓
9. 实验楼
10. 学院楼

总平面图

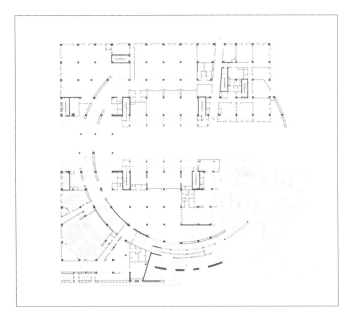

图书馆首层平面图

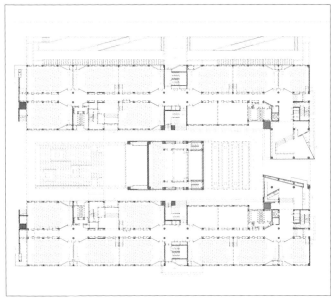

综合楼首层平面图

北京邮电大学沙河校区 · 2012—2014 · SHAHE CAMPUS OF BEIJING UNIVERSITY OF POSTS AND TELECOMMUNICATIONS

北京邮电大学沙河校区 · 2012—2014 · SHAHE CAMPUS OF BEIJING UNIVERSITY OF POSTS AND TELECOMMUNICATIONS

创造有自然和人文记忆的场所
A place with natural and human memories

清华大学深圳国际研究生院一期　TSINGHUA SHENZHEN INTERNATIONAL GRADUATE SCHOOL, PHASE 1
设计 Design 2018~2019　·　竣工 Completion 2023

地点：广东深圳　·　用地面积：23 054平方米　·　建筑面积：160 833平方米
Location : Shenzhen, Guangdong　·　Site Area : 23,054m^2　·　Floor Area : 160,833m^2

合作建筑师：王超若、闫晓婷、吕翔、张禹茜、董烨程、魏宏建
Cooperative Architects : WANG Chaoruo, YAN Xiaoting, LYU Xiang, ZHANG Yuxi, DONG Yecheng, WEI Hongjian

策　略：围合、借势、叠合、共享

摄影：存在建筑摄影
Photographer: Arch-Exist

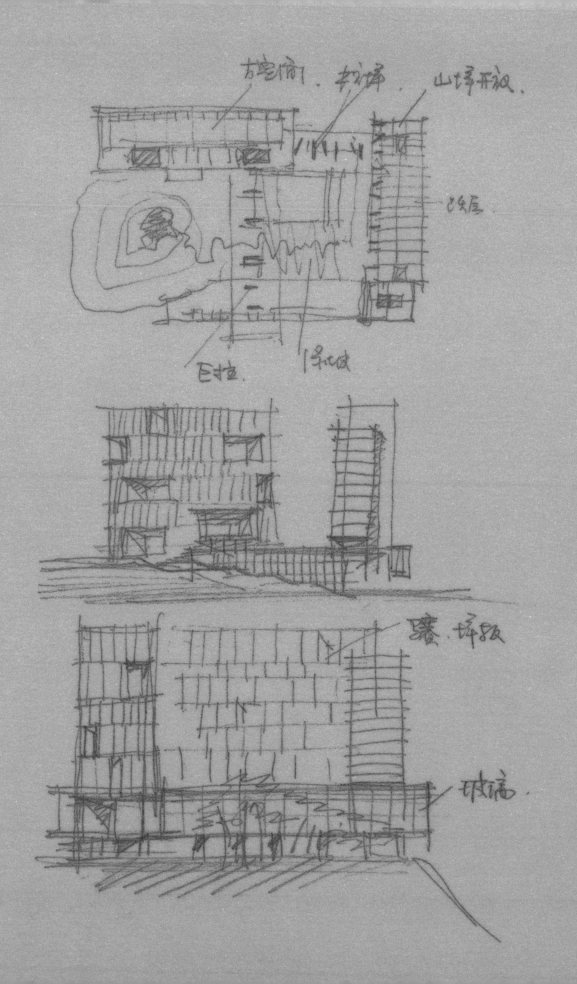

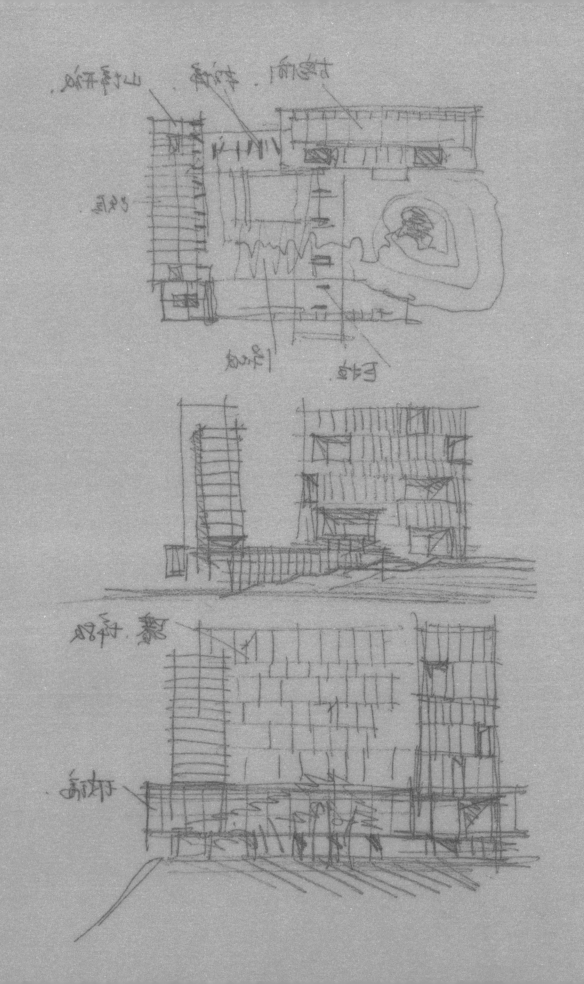

清华大学深圳国际研究生院一期是以教学科研功能为主的，集图书阅览、实验室、行政、会议及生活配套一体的立体高校教育综合体。用地位于深圳南山区大学城西校区，容积率5.4，建筑高度100m。建设用地局促，且近一半用地存在现状山体，建设面积指标高，无法以传统校园模式进行规划布局，同时还需满足国际化办学标准、高技术科研平台和多维度教学空间的要求。设计整合科研平台、教学空间、交流空间等功能模块，采用搭接、穿插和整合的方式，结合通风、采光、交通、绿化等要素，形成与山体共同交织、拥有多层次交流平台的"立体校园"。将绿植茂盛的山体最大化保留，与建筑巧妙结合成为设计的源点，同时与大学城环境融合为一体，形成"依山就势，望山理水"的绿色校园。以清华校园经典的"红区"作为基调，将有历史感的红砖建筑环境与岭南植被景观相结合，塑造百年传承的校园文化。空间布局融汇国际化学院特色和学科交叉教学理念，为师生提供交流互动、创新实践平台。

Phase I of Tsinghua Shenzhen International Graduate School is a multi-dimensional college education complex mainly serving for teaching and scientific research. Located at the western campus of Nanshan District College Town, it was faced with various challenges including limited site area and highly demanding design requirements, so a non-traditional design approach must be adopted to build a high-tech scientific research platform that meets international standards. The design has integrated modules including scientific research platforms, teaching spaces and communication spaces, where ventilation, daylighting, transportation and plantations were all taken into considerations to form a multi-dimensional campus that exists in harmony with the hills. On the site, the hills were preserved to the maximum extent to present a green campus, where red bricks of Tsinghua University were applied in this graduate school to compliment the local landscape features.

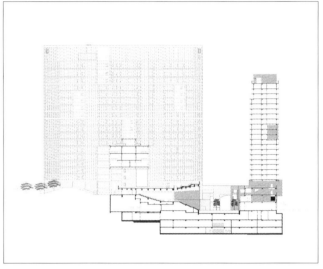

剖面图

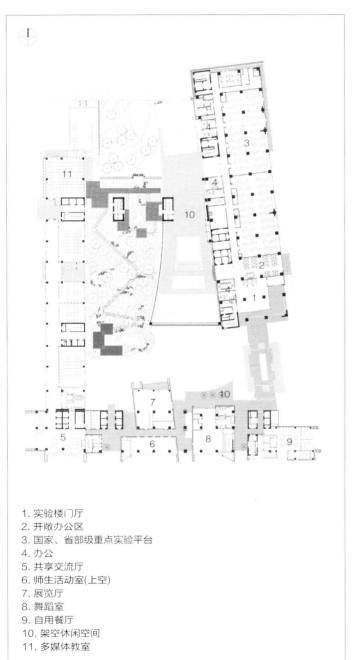

1. 实验楼门厅
2. 开敞办公区
3. 国家、省部级重点实验平台
4. 办公
5. 共享交流厅
6. 师生活动室(上空)
7. 展览厅
8. 舞蹈室
9. 自用餐厅
10. 架空休闲空间
11. 多媒体教室

四层组合平面图

清华大学深圳国际研究生院一期 · 2018—2023 · TSINGHUA SHENZHEN INTERNATIONAL GRADUATE SCHOOL, PHASE 1

LAND-BASED RATIONALISM III 253

清华大学深圳国际研究生院一期·2018—2023·TSINGHUA SHENZHEN INTERNATIONAL GRADUATE SCHOOL, PHASE 1

在高层建筑中关注人的尺度
Creating human scale in tall buildings

深圳万科云城 · SHENZHEN VANKE CLOUD CITY
设计 Design 2014~2016 · 竣工 Completion 2019

地点：广东深圳 · 用地面积：15 600平方米 · 建筑面积：178 000平方米
Location : Shenzhen, Guangdong · Site Area : 15,600m^2 · Floor Area : 178,000m^2

合作建筑师：邢野、金爽、曾瑞、张家博、吕柱、张胜强、李祥柱、燕立欢、刘鑫年
Cooperative Architects : XING Ye, JIN Shuang, ZENG Rui, ZHANG Jiabo, LYU Zhu, ZHANG Shengqiang, LI Xiangzhu, YAN Lihuan, LIU Xinnian

策 略：集约、混搭、开放

摄影：李季
Photographer: LI Ji

项目位于大规模城市开发的中心部分。该区域城市设计旨在打破以往分区明确的功能主义规划模式，探讨基于整体开发模式的小街区空间格局，并为派生的多样公共空间和步行系统留有条件。由政府主导、工作坊+集群设计的创造性组织形式为这次城市实验提供了运作基础。

功能混合导致了多样而复杂的边界条件，设计必须面对诸多问题，如城市主干道噪声、相邻300m高摩天楼带来的紧张感、3m的地形高差等，并有效利用相邻绿地资源，创建与周边酒店、商业等之间的积极关系。城市密度和交通强度的差别形成了由北向南、由西向东的压力场，决定了建筑的布局方式。

出于采光考虑，用地南侧布置U形平面公寓，自身围合出一个花园。北侧产业办公分解成150m高塔和L形100m低塔两部分，高塔以短边面对北侧摩天楼，低塔北边主动向南后退，策略性地回避摩天楼带来的压力，两部分围合出一个为四个地块共享的街角广场。L形低塔和公寓又围合出内部广场，通过底部架空，广场、花园以及周边绿地共同形成连续的公共空间，公寓花园和内部广场通过高差来实现领域划分。所有的设计策略都旨在于单体层面上积极践行最初的城市设计原则。

As part of a large-scale urban development project, the urban design of the region aims to explore a spatial layout of small blocks instead of the functionalist model, while reserving conditions for diverse public spaces and pedestrian systems. The project has adopted a creative way of organizing, where workshop and public design are combined with the government playing the leading role, laying the operational foundation for this experimental urban design practice.

Due to the complex functions, the design was challenged by noise from main roads, the pressure posed by an adjacent 300-meter-high skyscraper, the 3-meter altitude difference of the site, and the demand of creating a positive connection with surrounding hotels and commercial spaces.

An apartment building with a U shape plan is located at the south end of the site to optimize its daylighting conditions. Office buildings at the north consists of a 150-meter tower and a 100-meter L-shaped lower tower, which enclosure a shared square for the street corner, and another square is enclosed by the L-shaped tower and the apartment building. All design strategies are targeted for the implementation of the initial urban design principles.

深圳万科云城·2014—2019·SHENZHEN VANKE CLOUD CITY

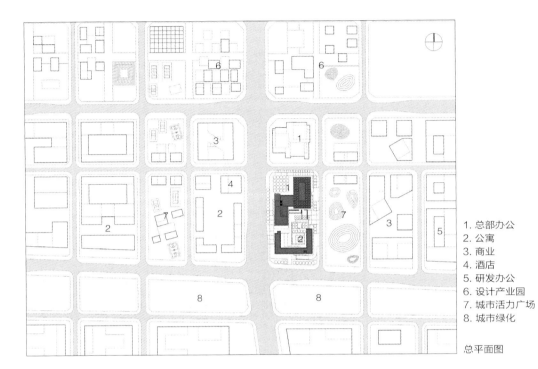

1. 总部办公
2. 公寓
3. 商业
4. 酒店
5. 研发办公
6. 设计产业园
7. 城市活力广场
8. 城市绿化

总平面图

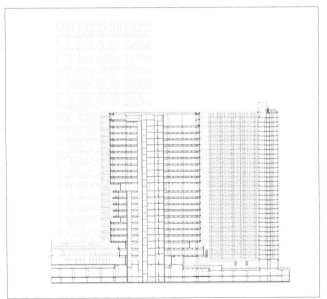

剖立面图

深圳万科云城 · 2014—2019 · SHENZHEN VANKE CLOUD CITY

流动的巨构
A giant structure of flow

济南舜通大厦 · JINAN SHUNTONG BUILDING
设计 Design 2017 · 竣工 Completion 2020

地点：山东济南 · 用地面积：17 307平方米 · 建筑面积：164 964平方米
Location : Jinan, Shandong · Site Area :17,307m² · Floor Area : 164,964m²

合作建筑师：任祖华、王庆国、朱巍、单晓宇、冯君
Cooperative Architects : REN Zuhua, WANG Qingguo, ZHU Wei, SHAN Xiaoyu, FENG Jun

策　略：集约、连续性、隐喻

摄影：张广源
Photographer: ZHANG Guangyuan

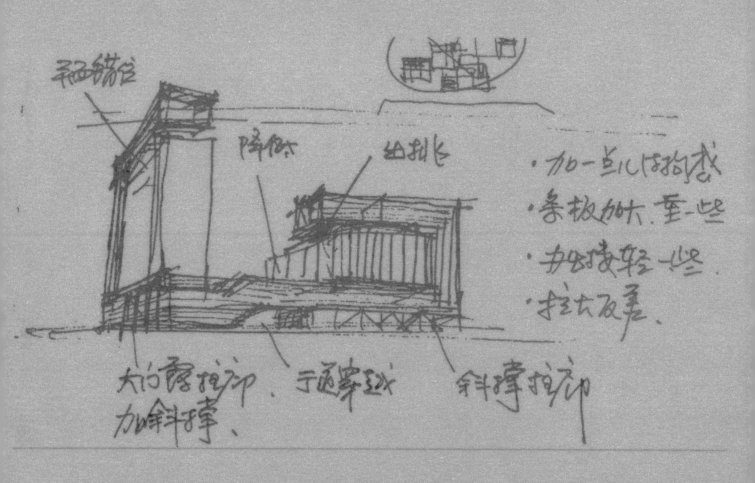

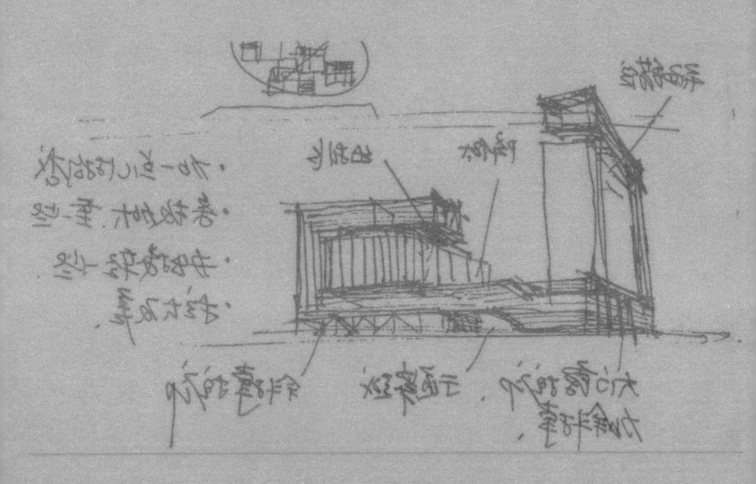

　　舜通大厦位于济南CBD东南角，该区域小街区、连续街墙的规划模式有利于形成连续的步行城市界面，但也使本来需要紧密联系的济南轨道交通集团总部办公楼和运营20条地铁线路的控制中心只能分别设在两个独立的地块内。设计遵循原城市设计主要理念，在首层分别形成两个地块连续的开放界面，创造骑楼空间，服务于市民穿行。场地东南角结合地铁出入口设置下沉广场，在有限的用地范围内无缝接驳城市公共交通。在三层以上用一组连续形体将两栋建筑整合为一体，实现功能连接的同时，也创造了有别于周边独立塔楼的城市空间。连接体的屋顶形成绿意盎然的立体花园，为员工提供多样化的休息空间，也成为周边高层俯瞰景观。源自地铁特有的窗型元素提示建筑自身的功能特色和个性，街角处地铁展览馆的巨大景窗形成了城市的视觉中心，以开放的姿态欢迎市民进入其中。

At the southeast corner of Jinan CBD area, where SHUNTONG Building is located, offices of Jinan Rail Transit Headquarters and the control center for 20 subway lines are separately located in two plots. Conforming to original urban planning concepts, the design has created open interfaces across the two plots and an arcade between the two plots. A sunken square is planned at the southeast corner of the site, where the subway entrance is located, to achieve direct connection of urban transportation. A series of continuous volumes have integrated the two buildings into a whole, creating a unique urban space sith a multi-dimensional garden on the roof. Windows with subway features have highlighted the unique feature and function of the building, and a huge window of the subway exhibition hall at the street corner poses a welcome gesture to citizens.

济南舜通大厦 · 2017—2020 · JINAN SHUNTONG BUILDING

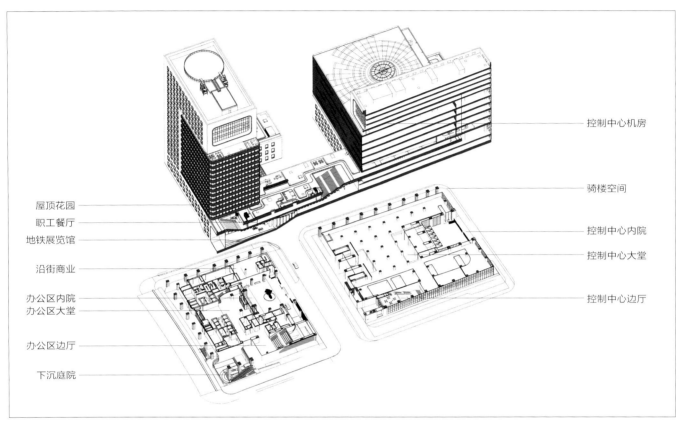

控制中心机房
骑楼空间
控制中心内院
控制中心大堂
控制中心边厅

屋顶花园
职工餐厅
地铁展览馆
沿街商业
办公区内院
办公区大堂
办公区边厅
下沉庭院

剖轴侧图

1. NCC数据中心
2. 管理办公
3. 企业展厅
4. 地铁展览馆
5. 出租办公
6. 咖啡厅

二层平面图

济南舜通大厦 · 2017—2020 · JINAN SHUNTONG BUILDING

LAND-BASED RATIONALISM III 267

合作激出海上花
Flower on the sea

青岛上合之珠国际博览中心 · QINGDAO SCODA PEARL INTERNATIONAL EXPO CENTER
设计 Design 2022 · 竣工 Completion 2022

地点：山东青岛 · 建筑面积：168 846平方米
Location : Qingdao, Shandong · Floor Area : 168,846m²

合作建筑师：关飞、董元铮、毕懋阳、张翼南、郭一鸣、盛启寰、薛强、王梓淳、付轶飞
Cooperative Architects : GUAN Fei, DONG Yuanzheng, BI Maoyang, ZHANG Yinan, GUO Yiming, SHENG Qihuan, XUE Qiang, WANG Zichun, FU Yifei

策　略：共享、开放、隐喻

摄影：存在建筑摄影
Photographer: Arch-Exist

项目主要功能为会议中心及上合组织国家的国别会展博览厅，选址青岛胶州如意湖入海口北岸，西临上合行政中心中轴线，东望跨海大桥，南望小珠山。作为一座以外交合作为主要定位的会议会展建筑，设计顺应宏大的自然山水格局，以生命与自然作为建筑的讴歌主题，以"海"和"珠"的意象，打造一个具有当代精神的会议会展综合体。纯洁抽象的白色、象征平等合作的圆形几何布局，贝壳状浪漫起伏的屋顶，都围绕同一个审美意象层层展开，力求体现滨海特色和当代中国开放与包容的时代形象。建筑采用大跨度钢桁架结构、铝板装饰TPO防水屋面和索网张拉玻璃幕墙，技术策略恰当、功能流线完备、建构细节简约，在短设计周期及快速建造的压力下高质量完成了这座标志性建筑。

Mainly consisting of a conference center and the SCO national convention & exhibition hall, SCODA Pearl International Expo Center is located at the north bank of an estuary in Qingdao. It borders the axis of SCO Administrative Center, with a cross-sea bridge to its east and Xiaozhu Hill to its south. The design conforms to the grand natural landscape and creates an image of "sea" and "pearl", showing respect for life and nature with the convention & exhibition complex with modern spirits. The building's circular layout, white color and the unrolling shell-like roof are all symbolizations of China's openness and inclusiveness in the new era. Appropriate technological strategies, well-designed functional circulation and delicate constructional details have guaranteed the high quality of the landmark.

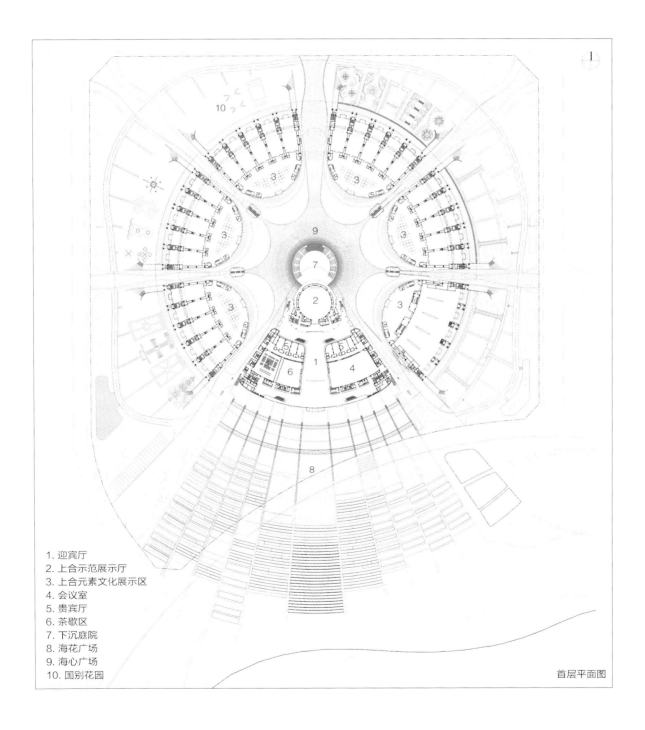

1. 迎宾厅
2. 上合示范展示厅
3. 上合元素文化展示区
4. 会议室
5. 贵宾厅
6. 茶歇区
7. 下沉庭院
8. 海花广场
9. 海心广场
10. 国别花园

首层平面图

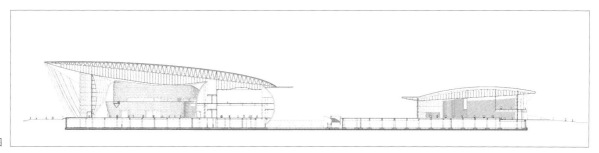

剖面图

青岛上合之珠国际博览中心 · 2022—2022 · QINGDAO SCODA PEARL INTERNATIONAL EXPO CENTER

青岛上合之珠国际博览中心 · 2022—2022 · QINGDAO SCODA PEARL INTERNATIONAL EXPO CENTER

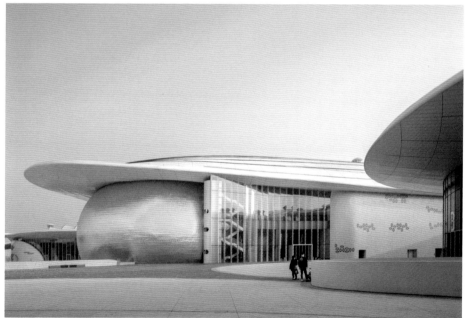

青岛上合之珠国际博览中心 · 2022—2022 · QINGDAO SCODA PEARL INTERNATIONAL EXPO CENTER

关于风格　ABOUT STYLE

　　有些学者总质疑我的设计跳跃性很大，一会儿是这个样子，一会儿是那个手法，似乎找不到连续性和稳定的个人设计表达，或者说"大"一点，没有什么风格。我完全承认这个评价，但是我认为本土设计不会导向设计风格，尤其是个人风格。因为它的价值和逻辑来自于因地制宜、一地一策，这个立场是不变的，但是方法和建筑形式呈现是不定的。当然也应看到许多有个人风格的著名建筑师在具体项目设计中也会对场所做出必要的反应，而这个反应并不会影响其表现个人风格的设计语言的系统性。所以我又认为本土设计并不排除个人风格，甚至也可以用一个时尚的表述方式——"本土设计+X"，就是在本土设计的原则下，可以表现每个建筑师的个人喜好和独特的设计语汇。

　　而我本人对本土设计的悟道和对个人风格的回避是一种自在的态度，使我应对设计项目更从容、放松，不刻意加进个人化的语言，也不希望设计因为个人的语言使其在环境中凸显出来，让周边的环境成为配角和陪衬。实际上我觉得大多数职业建筑师因为工作状态、各种条件以及个人的能力所限，不会发展出明显的个人风格，但他们仍应该遵守"场地决定论"的规律去思考设计，因为这是一个基本的常识和逻辑。

　　本土设计的普适意义是建立在建筑与自然、人文环境关系的基本规律上，是一种基本理性的方法论。但之所以将这种近乎"公理"的东西当做一个方法论，与其他建筑理论体系并列，是因为在现实中有太多不遵从这个规律、不讲这个"公理"的设计。许多建筑师过度关注建筑艺术性、或趣味性、或个人风格；许多建筑师过分相信规范、标准和技术解决方案；许多建筑师屈从权力、资本的取向，不坚持理性客观的立场。凡此种种，都会忽视建筑与环境的关系，更不用说具备用这种特定的关系去创作特定建筑的能力，以及这种能力的培养和提升了。应该说，即便熟练地掌握了本土设计的理念和方法，能否做出适应环境的优秀建筑作品来也还是个未知数，"不出错、少出错"不意味就是优秀、就有创新性。因此本土设计在原则立场和策略方法之外，还有着广阔的创新空间，也希望有才华和条件的建筑师们在"做对"的前提下去追求个人的风格。

针灸式：乡村振兴的"末微介入"犹如针灸，扎一点串全身，以点带面逐渐更新。

保格局：乡村既有格局源自多种因缘生长而成，设计应尽量保持这种格局，在原址上更新，提升乡村的生活环境。

留乡愁：乡愁是乡村的历史留在村民中的记忆，更新中一定要小心翼翼地保护好有意味的历史痕迹，让乡村的文脉和情缘不断。

注活力：解决乡村空心化、老龄化，要通过吸引青年人回乡和城里人下乡重振生机。

兴文化：乡村传统文化的保持和传承，振兴新文化，使村民增加自信。

乡村振兴
RURAL REVITALIZATION

守护田园、留住乡愁是重振乡村的基础，也是复兴乡村文化的前提，扶助乡村要因地制宜。轻设计、陪伴式、保持和提升乡村特色、助力产业发展、帮扶弱势群体、推动乡村可持续发展是本土设计的策略。

让昆曲在水乡传唱
Haunting drama in water villages

昆山西浜村昆曲学社及乡村工作站
KUNSHAN XIBANG VILLAGE KUN OPERA SCHOOL &

设计 Design 2014 · 竣工 Completion 2016

地点：江苏昆山 · 建筑面积：2 868平方米
Location : Kunshan, Jiangsu · Floor Area : 2,868m²

合作建筑师：郭海鞍、沈一婷、向刚、冯君
Cooperative Architects : GUO Haian, SHEN Yiting, XIANG Gang, FENG

　　西浜村位于阳澄湖畔、绰墩山北，是昆曲文化的发祥地。如今村庄已经破落，为了恢复乡村中的昆曲文化，带动乡村复兴，将村中的四座老宅院修复改造成昆曲学社。设计通过粉墙和竹墙形成梅兰竹菊四院，结合水系搭建戏台，通过两层游廊的穿插，形成一个空间丰富、光影交错的昆曲研习之场所。

　　昆曲学社建成后取得了良好的社会文化效益，为了继续巩固乡村微介入规划的效果，在当地政府的支持下，我们将相邻的西南民宅改造成乡村工作站，以便提供长期伴随式的服务。通过屋面变化丰富的新建工作室与精心修复后的老房子相结合，既形成了满足新生活新工作的空间，又保留了传统房屋的生活空间，传承了过去，记住了乡愁。乡村织补过程中多采用地域建筑材料，并创新性使用了竹木结构、金属瓦、竹木板、雾化玻璃等新材料和新技术，希望为更多农房改造提供可借鉴的范例。

Located along Yangcheng Lake, Xibang Village is the cradle of Kun Opera culture. To rejuvenate Kun Opera and the village itself, the design has renovated four old courtyards into Kun Opera School, where opera stages were planned along waters in the courtyards. A two-story corridor is located in the school, creating a unique and attractive place to study Kun Opera.

After completion, the Kun Opera School has gained cultural and social benefits. To consolidate the good results of this sort of "micro-intervention", a rural working station is renovated from a folk house under the support of local government, so that the efforts of rural rejuvenation can be carried forward. Newly built working spaces are integrated with elaborately repaired old houses, where the past was inherited and nostalgia was remembered at heart. Local materials are applied with innovative technologies and constructional methods.

昆山西浜村昆曲学社及乡村工作站·2014—2016·KUNSHAN XIBANG VILLAGE KUN OPERA SCHOOL & RURAL WORKSTATION

针灸：借用中医针灸以穴位刺激经络的方式，在西浜村复兴中从局部的农房改建为昆曲学社为起点，激发整个乡村文化回归和有机更新。

艺术：以原竹格栅的疏密关系形成光影的变化，表现昆曲的节奏韵律，形成空间艺术和昆曲艺术的二重奏。

保格局：保持原有村落肌理和宅地、水巷的关系，用与传统相协调的材料和语言，将衰败的乡村聚落重新织补起来，表达了对乡村原有风貌的尊重和保护。

材料：尽管采用了当代的钢结构和铝门窗、金属瓦，但注重对传统当地材料的意向表达，近人的原竹和草泥墙保持了乡土的气息和亲切的触感。

色彩：粉墙、黛瓦、竹棚，在绿树掩映中呈现出江南典型的水墨色韵。

1. 南门房
2. 梅院
3. 多功能厅
4. 戏台
5. 菊院
6. 序厅
7. 竹院
8. 化妆间
9. 舞蹈教室
10. 教室
11. 教师办公
12. 兰院
13. 食堂
14. 古桥
15. 水面
16. 工作坊
17. 图书室
18. 卧室

首层平面图

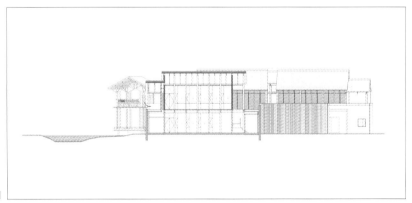

剖面图

昆山西浜村昆曲学社及乡村工作站 · 2014—2016 · KUNSHAN XIBANG VILLAGE KUN OPERA SCHOOL & RURAL WORKSTATION

昆山西浜村昆曲学社及乡村工作站 · 2014—2016 · KUNSHAN XIBANG VILLAGE KUN OPERA SCHOOL & RURAL WORKSTATION

昆山西浜村昆曲学社及乡村工作站 · 2014—2016 · KUNSHAN XIBANG VILLAGE KUN OPERA SCHOOL & RURAL WORKSTATION

LAND-BASED RATIONALISM III 289

窑光点亮振兴之路
Revitalizing road brightened by kiln light

昆山锦溪祝家甸砖厂改造 · KUNSHAN JINXI ZHUJIADIAN BRICKYARD RECONSTRUCTION

设计 Design 2014 · 竣工 Completion 2017

地点：江苏昆山 · 建筑面积：5 154平方米
Location : Kunshan, Jiangsu · Floor Area: 5,154m^2

合作建筑师：郭海鞍、张笛、沈一婷
Cooperative Architects : GUO Haian, ZHANG Di, SHEN Yiting

策 略：针灸、轻介入、乡村营造

摄影：张广源
Photographer: ZHANG Guangyuan

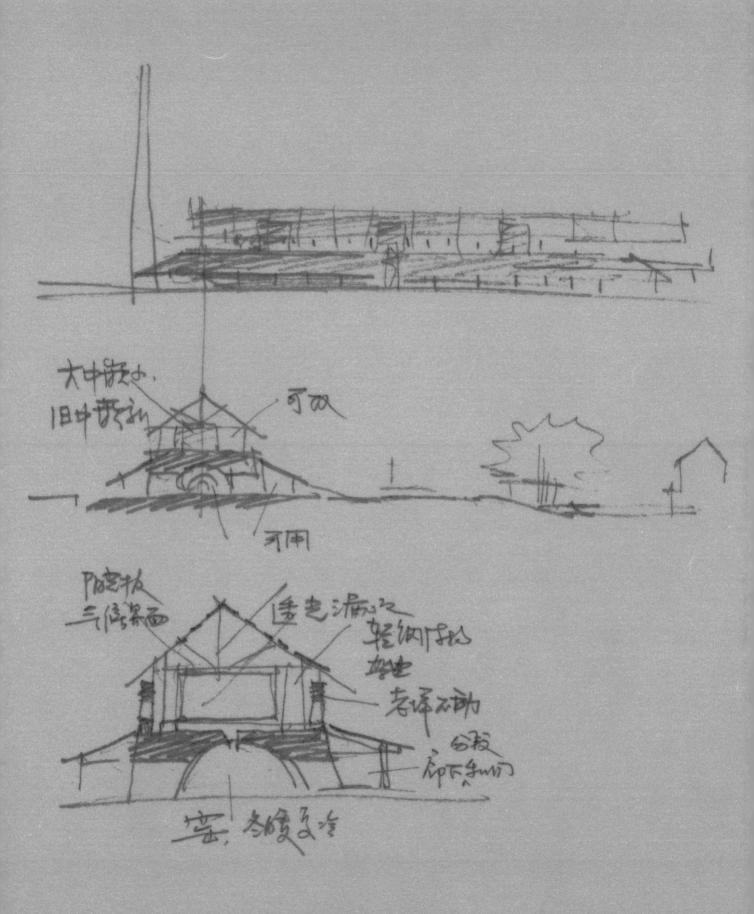

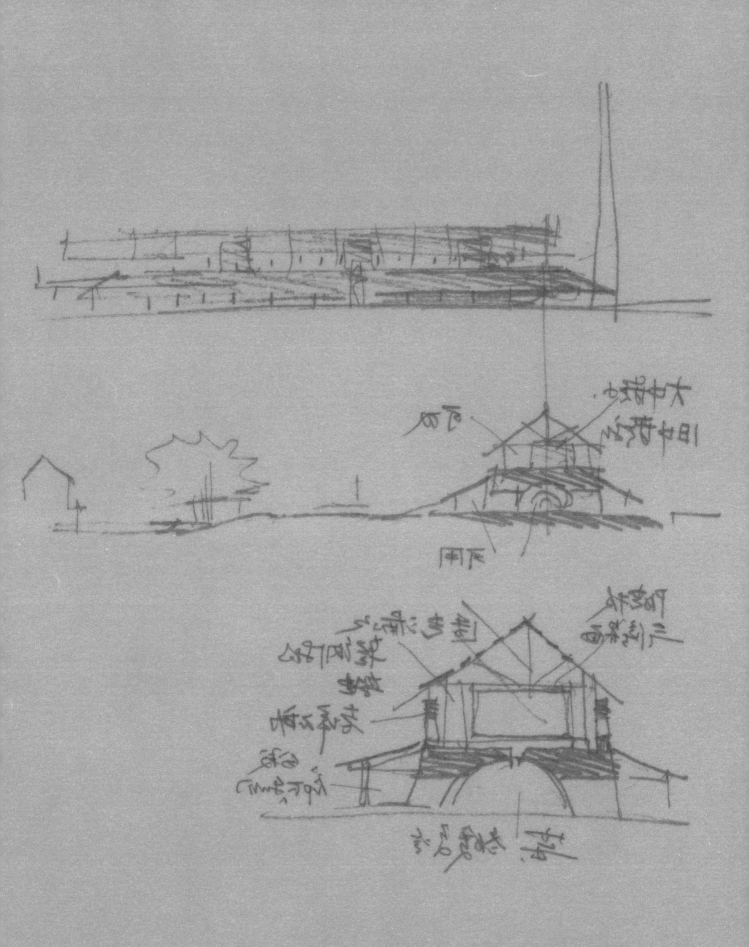

祝家甸是古代宫廷所用金砖产地，至今仍有古窑在继续使用，但村落已破旧不堪。乡村复兴计划采用"微介入"的方式，将一座建于20世纪80年代的废弃砖厂改造为古砖博物馆，容纳展厅、手工坊、咖啡吧等空间。阳光穿过残破的屋顶投入室内的动人场景被设计者以透明瓦延续下来。室内空间采取模块化设计，所有地板单元、家具单元、设备单元、镂空单元均可移动和替换，为小建筑提供更多的可能性。为了将底层结构加固的钢拱融入老窑的氛围，在钢拱上增加精美的砖拱，形成的效果如同一条时空隧道，可以回味当年热火朝天的生产场景。入口处独立基础的钢楼梯将老坡道保护其中，新老交叠而存。建筑山墙上"淀西砖瓦二厂"的旧水泥字，则保留着近半个世纪的乡愁和回忆。

Zhujiadian Village was once a place producing gold bricks for the imperial family, with several kilns in use till today. Through "micro-intervention", a deserted brickyard was renovated into a museum with an exhibition hall, a DIY workshop and a coffee shop. The transparent tiles have preserved the appealing scene of sunlight pouring in through the cracks of the roof. All the components of the interior space are designed with modules that allows for mobility and replacement. A brick arch was built over the steel arch, reminding people of the history. A steel staircase at the entrance covers the old ramp to present the co-existence of the old and the new. On the wall, old cement-casted name of the brick factory was preserved.

昆山锦溪祝家甸砖厂改造 · 2014—2017 · KUNSHAN JINXI ZHUJIADIAN BRICKYARD RECONSTRUCTION

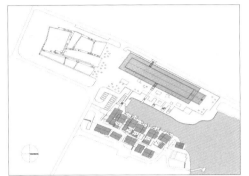

总平面图

改造前的废弃砖厂

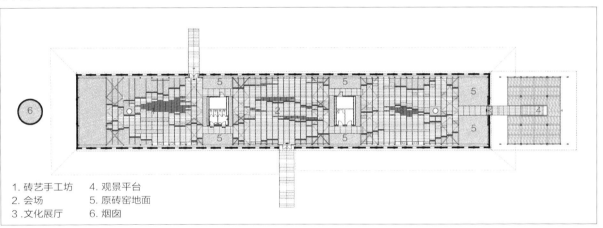

1. 砖艺手工坊　4. 观景平台
2. 会场　　　　5. 原砖窑地面
3. 文化展厅　　6. 烟囱

首层平面图

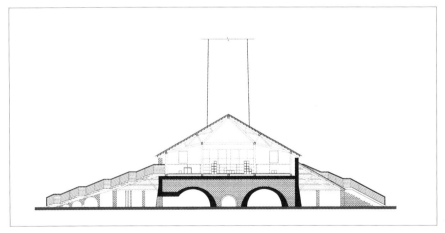

剖面图

LAND-BASED RATIONALISM III *293*

昆山锦溪祝家甸砖厂改造 · 2014—2017 · KUNSHAN JINXI ZHUJIADIAN BRICKYARD RECONSTRUCTION

废墟变佳处
Rebuilding scenic spots in an abandoned place

昆山小桃源 • SMALL LAND OF PEACH BLOSSOMS IN KUNSHAN
设计 Design 2017 • 竣工 Completion 2021

地点：江苏昆山 • 建筑面积：6 989平方米
Location : Kunshan, Jiangsu • Floor Area: 6,989m²

合作建筑师： 郭海鞍、冯君、沈一婷、孟杰、向刚
Cooperative Architects : GUO Haian, FENG Jun, SHEN Yiting, MENG Jie, XIANG Gang

策 略：保留、更新、融合、意境

摄影：徐晓飞
Photographer: XU Xiaofei

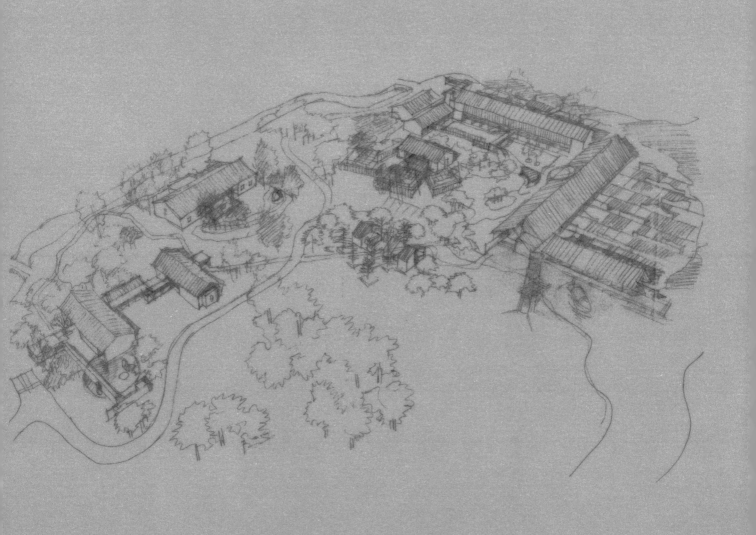

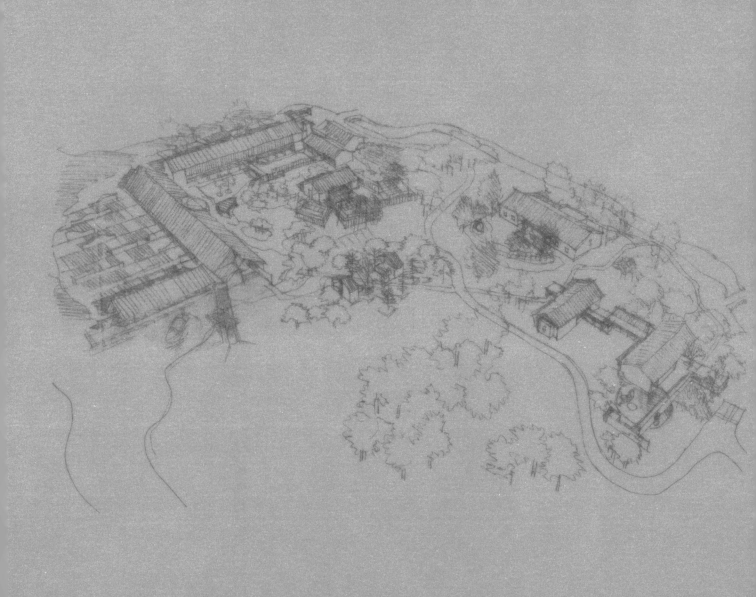

小桃源，是以《玉山雅集》意境再现为主题的昆山文化复兴基地，也是一场基于乡土文化传承的城乡更新实践。这里原本是一处早已废弃的麻风病院，改造工作最大限度地保留了树木和湿地，小心翼翼地修复原有建筑，并按照《玉山雅集》的诗词加以改造，形成拜石坛、寒翠所、来归轩、可诗斋、听雪斋、芝云堂、钓月轩、秋华亭等多处景致。对历史情境的文本解读，展现了一座有故事的、美好的、生机盎然的新时代江南园林及田园建筑。

The Small Land of Peach Blossoms is a base for Kunshan Culture Rejuvenation, as well as a project of urban & rural renovation based on rural culture inheritance. Renovated from a deserted leprosy hospital, the project preserved a large portion of trees and wetlands, and carefully repaired existing buildings. Furthermore, a series of buildings and pavilions are built in accordance poems in the book of *Yushan Yaji*, interpreting historical scenes while presenting a vigorous and beautiful image of Chinese gardens in the new era.

昆山小桃源 · 2017—2021 · SMALL LAND OF PEACH BLOSSOMS IN KUNSHAN

1. 玉带桥
2. 拜石潭
3. 寒翠所
4. 来龟（归）轩
5. 秋华亭
6. 可诗斋
7. 钓月轩
8. 听雪斋
9. 芝云堂
10. 码头
11. 水塔观景塔

首层平面图

改造前原状

昆山小桃源 · 2017—2021 · SMALL LAND OF PEACH BLOSSOMS IN KUNSHAN

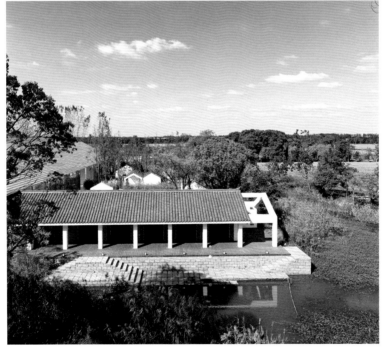

LAND-BASED RATIONALISM III 303

昆山小桃源 · 2017—2021 · SMALL LAND OF PEACH BLOSSOMS IN KUNSHAN

LAND-BASED RATIONALISM III *305*

设计随笔

2013年我应邹德慈院士的邀请参加了中国工程院重点课题"村镇规划与治理",具体负责"村镇风貌"一章。自那时起,我带领郭海鞍杜儿位硕研究生开始了乡村的研究,调查走访了不少乡村,也去中国台湾地区和日韩等国考察学习,逐步对乡村文脉的保护和传承有了一些认识,而不像以往大家仅仅关注表象的风貌问题。2014年,昆山市的有关领导听说了我们的乡村研究工作,主动邀请我们去昆山的几个村子看看如何开展乡村整治和振兴的工作。

位于昆山西部的西浜村地处阳澄湖和傀儡湖之间的河滩地带,村子规模不大,不新不旧的农居沿河汊分布,房前种菜,房后拴船,岸边大树掩映,水中鸭子成群,很有水乡的特色。之前政府曾计划将村庄整体拆迁,让出用地打造水乡特色小镇;而我提出可否原状保留,不大拆大建,就利用部分闲置破败的宅院做微更新改造,这样不仅可以降低开发成本,也利于蟹农的劳作生活,在此基础上发展民宿和田园旅游,应该是游客们更感兴趣的。另外在调研中得知,昆山就是昆曲的发源地,元代一位贤士顾阿瑛曾在村中生活,著有一册《玉山雅集》,记载了他与朋友演习昆曲的二十四佳处。这让我们找到了设计的思路,通过村落及周边田园景观的微改造,逐渐再现二十四佳处,让昆曲的文脉带动乡村文化的振兴。

最先入手的是昆曲学社,利用一处河边已倒塌的宅基地,按照原来的格局复建的一组院落式小建筑。使用功能是请昆曲演员来办小昆班,定期在这里教村里或邻近社区的小朋友学唱昆曲。设计强调"传承出新",用传统农居的语汇和钢结构、竹栅帘、抹泥白墙构建灵动的空间和院落,入口大门掩映在竹林菜地之后,低调而雅致。河边搭一竹楼作为戏台,砖砌的观众席在小河对岸的大树下,一座古石桥复建在旁边,营造出昆曲实景表演的空间场所,轻松而有趣。听说每年都有200天左右的培训演出活动,昆曲真的还乡了。

昆曲学社挨着一处农宅,住着一位养鸭子的老奶奶。前几年老人去世,家人将宅子卖给了城投公司,我们便又将其改造成了一座小小

的乡村工作站。改造中我们特意加固保留了老墙老顶让时间的痕迹留下来,将小院加棚变为大厅,门外加敞廊存放工具和自行车,窗前围上镂空矮墙让内外窥望,加上简朴实用的装修陈设,这里成了我和团队常常落脚的好地方。早晨在鸡鸣中醒来,去稻田河边散步,鸟鸣鸭叫,微风中散发着泥土的芬芳,让我们对乡土又多了一分感情。

昆山的南部是锦溪,长白荡湖边有个烧砖的小村叫祝家甸,据说从明朝年间就有烧砖的产业。工匠以湖泥为料烧出来的砖细润质坚,敲起来有如金属声,谓之金砖,是宫殿和庙宇大殿中铺地的上品。去村中看时,一排圆丘状的古窑多已破败,少数几座还冒着烟火,但烧制的砖质量也大不如前,出窑后装到停在旁边河汊中的小船运到浙江去加工销售。村中会烧砖的人也不多了,大多数年轻人去周边的工厂里务工,一排排整新过的农宅人气也不旺。我们看到村口有一座闲置的霍曼式砖窑,之前是烧盖房的红砖,因为环保和保护农田的政策早已停产,但留下的窑体和煤棚空间倒是可以利用。我们整修加固了煤棚轻钢结构,沿用原来的红色机瓦覆盖屋顶,借用现状瓦破漏光的样子,将部分改为阳光板透光瓦,解决了内部采光的问题;也是利用阳光板在内部形成了闭合的界面,形成了有空调的会议接待用房,而大部分空间仍维持原来的开敞性不用能;旧墙旧窑顶都保留原样不动,窑道内部局部加钢拱防护,窑外侧棚廊改为茶室、展厅;向湖端部分

延长屋架和平台便于观景；小河滩对岸又做了一组民宿型精品酒店生意一直不错。游客来了，村里的年轻人也回来了；孩子们来学砖艺了，家长们就带他们去村子那头看古窑；游客进了村，村里的农家乐也开起来了；再后来会有做陶做砖雕的作坊在村里落户；再后来村里的人气便会旺起来，这就是我们用"针灸疗法"在乡村的振兴计划。

在西浜村的南面，城市道路旁有一片密密的树林，树林中隐藏着一个破落的大院，几座简陋的砖房散落其中，据说这是一个建于20世纪50年代的麻风病院。这种可怕的传染病无药可治，患者被集中关在病院中自生自灭。当然这种病在当地早已绝迹，这个病院也无人再用，废弃在林中。第一次走进这座阴森的院子，还似乎有些恐怖，但唯有树林越长越密，野草藤蔓噬食着陋屋内外，自然之物重新占据了空间。由于这里靠近水源保护地，原则上是不允许再建设的，但看到这些即将成为建筑垃圾的废墟，像个垂死的老人，能不能去扶他一把？设计的策略还是"以留为主适当增补"，调整布局下手要轻，景观上尽量保留已成林的树木，结合《玉山雅集》，依据顾阿英当年对各个佳处的文字描述营造主题场所，希望改造成年轻人喜欢的生态体验基地。建成后听说反响不错，成为政府领导接待客人、介绍昆山田园生态环境和乡村振兴的场所。

其实，以昆山实践为起点，我们这支乡建小团队已经在江苏、浙江、福建、山东等地开展了更多的实践活动。每次看到他们的设计成果都十分欣慰，当然也会指出问题提些建议，出些管用的点子，乐此不疲。乡村振兴是个长久的工作，需要陪伴。

评论·访谈

REVIEWS & INTERVIEWS

厚土重本　大地文章
——崔愷和他的"本土设计"
ARCHITECT CUI KAI AND HIS "LAND-BASED RATIONALISM"

金秋野

崔愷在《本土设计Ⅱ》的标题文章"关于本土"中选用了一幅插图，那是一株枝繁叶茂的大树，根部深入土壤，由社会政治经济资源、地域文化资源、科学技术资源、工艺材料资源、气候资源和土地环境资源构成，共同撑起主干——"立足本土的理性主义"(land-based rationalism)，而后发展出地域建筑、文脉建筑、社会建筑等枝叶。此图解应看作崔愷自2008年提出"本土设计"观念并出版同名专著后，对个人建筑观的一次反思、细化、再组织的结果。针对质疑，崔愷指出"本土设计"若干要点，如立足现实，避免概念化操作；重视土地，解决具体问题；发展，而不是停滞或回归；代表一种立场或文化策略，而不是一种主义或风格，故称之为"本土设计"而非"本土建筑"。相比业界的众说纷纭，书末几篇理论文章更能引人思考。"本土"引出的话题，包括"乡土""地域""传统""国家""土地""自然"等，都得到系统探讨。有意思的是，人们对"本土设计"的看法，往往因不同的价值取向而不同。如德尔夫特大学建筑学院前院长尤根·罗斯曼(Jürgen Rosemann)即认为崔愷的理论和实践须放在"批判地域主义"的语境下讨论，因此有别于主张复古的保守主义建筑师们。他因此称崔愷为"现代的传统主义"。而学者朱剑飞在仔细分析了崔愷所代表的国营设计院在中国建筑界中的地位和作用后，指出崔愷努力平衡社会实用价值和个人职业追求之间的矛盾，是"和谐"这一传统观念在现代建筑中的具体表现。他梳理了崔愷设计的五原则，指出这与建立在西方文化语境下的"批判地域主义"并无关系，是中国式的世俗灵活性和时代开放精神的综合体现。他因此称崔愷为"抽象现实主义"。

我同意朱剑飞的部分观点，即崔愷的设计思想与批判地域主义无关。当年亚历山大·楚尼斯小心地用"批判"(critical)来限制附丽于"地域主义"之上的浪漫乡愁，借芒福德来强调"批判地域主义"价值观中显著的理性态度、与国家民族主义的区别，但终究是建立在同"全球化"的对抗之上的，相貌中带有法兰克福学派的愁容。崔愷并没有在任何方面自外于现代性或现代主义，他对本土或历史的理解是基于"现代世界"这个既成事实之上的，他想通过建筑表明的是，一个现代中国建筑师，如果同时具有深沉的本土意识和广博的世界胸怀，现代也可为我所用，传统亦可回馈于现代，这无疑也是"现实理性"的一部分。

可是，尽管朱剑飞指出崔愷在对待本土文化、自然环境、历史遗产和现实生活方面一概给予足够的尊重，却并未重视"本土设计"观中的一个重要内容，即作为一种设计策略所体现出的自觉的国家意识。换句话说，崔愷的"本土"之所以不同于泛指的"乡土"或"地域"，是因为它特指就"中国本土"。"乡土"或"地域"概念的对立面是"全球化"或更广泛意义上的"资本主义"，背景是以新马克思主义为基础的西方现代文化批评；"本土"概念的对立面则是"无根"，特指现代文化冲击下近于失语的中国知识界（包括建筑设计领域），和城市文明侵袭下几乎失去生存想象的中国民间社会。"本土"观的核心在"厚土重本"，扎根现实，培育根系，这本是异常艰难的工作，故崔愷无数次在报告和文章中提到"历史责任感"，这是中国的事情，中国人不来关心谁来关心？这与全球化时代的世界主义并不抵触也不狭隘，更不同于西方知识界深以为戒的国家民族主义。我们不妨更进一步：完整的国家意识岂止是一些中国建筑师的主动追求，甚至可以视为他们的道德准则，在整个文化风气趋向个人主义的时代，他们更像是社会中努力延续传统的支柱力量，而不只是被动的跟随者。这份国家意识正是中国传统知识人的核心价值，与个人主义的"优异性"追求相比，它以个体为单位自发定义着"现代中国"的内涵，更能让中国人找到在世间的位置，同时也可以完全是批判性的、具有独立意识的。以这种平稳的坚持，崔愷显示出中国建筑师的独特性。

崔愷的作品之所以可称"本土"，还有个原因就是大部分项目都在国内。偶有例外，也是代表国家形象的驻外建筑。在中国驻南非使馆项目说明中，崔愷谈到："建设使馆，既要在异国地域中表现中国的精神，也要尊重所在国的文化，更要与邻里和谐相处。"建筑是大国重器，不只是表达个人风格的载体。回顾工作室长长的作品名单，不乏大量具有深厚本土文化特征的项目，如北京首都博物馆(2005)、安阳殷墟博物馆(2006)、辽宁高句丽遗址博物馆(2008)、敦煌莫高窟游客中心(2014)、泰山桃花峪游客中心(2010)、玉树康巴艺术中心(2014)等。与文化潮流相应，现代建筑学的知识话语中，容许充分个性表达的小型项目（如别墅或私人博物馆）受到关注，代表宏大叙事和国家文化的项目反而被忽视。崔愷给自己提出的一个问题是：如何在大型项目和学科发展之间重建关联？与小型项目相比，大型项目各方面的复杂程度呈几何级增长，附带的象征意味更让人难以招架，崔愷的解决方案是柔性的，即将设计师的表达欲放至极低，让具体的自然条件和历史人文信息成为形式驱动力，这是"本土"的第二层含义，即主动适应客观条件，力求综合，反对偏颇，避免平庸。建筑师成为环境的协调师，任务是激励和修补，让有缺陷的现实趋于完整。有时这种修补甚至延伸到城市设计层面，对区域环境有所交代。通过从场地中提取"客观"的形式要素、让环境说话，设计师避免了本人和业主双方主观的形式偏好和观念冲突，也给历史元素、地方风格，给现代建筑理论和形式语言留出机会，让各种错综的因素，包括历史的、现实的、经济的、政治的、技术的、人文的、普罗的、精英的、城市的、自然的，在一个谦逊的形式解答中达到平衡。

以"批判"为目标的知识系统，驱动力永远都是人为的知识和造型，坑越挖越深，路越走越窄；以"平衡"为目标的知识系统，驱动力在于不断发展变化的外部世界，心态保持开放，视野是开阔的，语言不离常识。当然深与广之间也要平衡，我想若有一种健康的建筑观，它不仅应是纵向上观念批判的产物，也应是横向上谋求平衡的艺术。崔愷在很多地方谈到"现实理性"，它首先应是一种实践理性，本土、传统、自然、技术、国家、民族，都是实践的对象，永远变化、保持常新。为了平衡这些复杂的要素，建筑师不能太强调个人实验，也不能预设概念、违背常识，是一种社会实践上的"允执阙中"，在这个过程中，任何一方的价值观都得到充分尊重，也并没有牺牲建筑师的职业操守。

在康德的表述中，实践理性与伦理息息相关。所谓实践理性就是事情发生的时候，如何判断、如何选择，它针对的是永远不断变化的外部世界，并不指望通过一套个人化的形式语言(风格)加以解决。现代建筑知识系统对先验和绝对形式的追求，显然是过多了，对环境的尊重多是口头上说说，建筑师都把带有强烈个性特征的、追求某种"本质"的形式语言看作职业桂冠。对崔愷来说，假如在面对具体问题的最佳伦理标准是"平衡"，那么须在审美上加以体现，成为沟通实践理性与纯粹理性的桥梁。在中国人眼里这不是什么新鲜事物，从认识的根源处看，"平衡"不仅可以是绝对的，且可以通过审美来体现，即中国艺术中对"度"或"分寸"的不懈追求。"度"成为"本体"，这在李泽厚的"历史本体论"中已有阐述。或许是在无意识中，崔愷触及本土哲学的关键问题，即"平衡"不仅可以是伦理的本体，也可以是审美的归旨，而它一直在变化中。重视"变化"、强调"和谐"是中华文明的核心价值，言外之意，大凡"观念"都追求终极而趋于静止、缺乏弹性而失于偏颇，较真起来冲突不可避免，现代世界的很多麻烦因此而起。避免冲突的最好方法就是彼此尊重、和而不同。

如果说崔愷的作品和思想体现了某种"本土"意识，也不是通过具象的形式符号表达出来的，而是宏观认知、判断选择上的一种分寸感，这也旁证了传统继承之难，因为它并不在可见的地方。相比之下，追求终极和静止，反而是容易形式化的。对变化的认知和对分寸的把握，并不纯然是中国特有的东西，但只在中华文明中，它被置于核心性地位，并在各个时代寻求新的表达。在这一过程中，传统知识人(士)不仅具有文化的解释权，也充当了知识更新中介人的作用，个人与国家、与传统一直是紧密相连的，这一关系却在现代中国几近瓦解，仅在有限范围内以特定的方式存留。应该说，崔愷在职业生涯中表现出的国家意识和历史责任感，正是本土知识人与国家传统关系模式的现代衍生物，它在现代社会中遭到强烈冲击，也只有依靠个人自觉的文化—历史归属感得以延续，代表一种现代建筑职业身份的"本土模式"，并在实践中自证合理。《论语·学而》有言：君子务本，本立而道生。崔愷的作品和思想，也会因其对本土意识的执守和对大地河山的热爱而自成文章。

——原载于《建筑学报》2016年第8期

面向中国本土的理性主义设计方法
——崔愷院士访谈
LAND-BASED RATIONALISTIC DESIGN METHODS:
AN INTERVIEW WITH ACADEMICIAN CUI KAI

崔愷 范路
对谈时间：2018年09月14日
对谈地点：中国建筑设计研究院本土设计研究中心

以往，大家对于崔愷院士作品和思想的讨论主要聚焦在"本土"上。但细读《本土设计 II》，尤其是江苏建筑职业技术学院图书馆项目和介绍其设计理念的《设计的逻辑》一文，会发现其中蕴含着深刻的理性主义建筑思想和设计方法。这种理性主义注重美学意境、生活体验、工程技术和经济合理的多元整合；强调设计意图充分贯彻、建筑语言丰富变化又逻辑一致的建筑品质；更是面向中国本土、传统智慧和绿色人文的可持续性思想方法。

范路：您的两本作品集，从2008年的《本土设计》到2016年《本土设计 II》，英文书名从"Native Design"改成了"Land-based Rationalism"。为何将"Design"换成了"Rationalism"，这是否意味着设计理念的变化？

崔愷：2008年《本土设计》出版时，英文书名简单直接翻译成"Native Design"。但后来觉得不够好，"native"容易被人误解成带有"本地的"和种族意味的含义，显得不够开放。所以到了出版《本土设计 II》时，我想把"本土设计"这个词的英文翻译改一改。朱剑飞老师帮我想了3个，我最后选定了"Land-based Rationalism"。因为我觉得我在创作中对现场环境的分析是比较理性的，而不是那种笼统的概念和牵强附会的隐喻。2018年，我在伯克利大学跟弗兰普顿（Kenneth Frampton）老先生交流的时候，送了他这本书。他说以前不太了解中国的设计院，但现在发现中国的设计院也会有好的设计师，比如说我的作品和"Land-based Rationalism"，都挺有意思。当然，也有学者见到标题后会说，你这个"理性主义"（Rationalism）是不是跟包豪斯等欧美建筑思想有关系。

范路：在现当代建筑学语境中，理性主义是个内涵丰富的概念，狭义来说主要指1920、1930年代以包豪斯、柯布西耶、特拉尼等为代表的建筑思潮，指向与功能主义、机器美学、国际风格等相关的建筑原则。往后延伸，它也包括1960年代以罗西为代表的强调人、建筑与城市整体性的新理性主义。而从广义上来看，理性主义意味着现代建筑学最重要的思想方法。它可以追溯到西方古典主义、文艺复兴的传统，并经由18、19世纪的欧洲启蒙运动发展至现当代，是强调基于科学精神和理性分析的建筑原则与设计方法。

细读《本土设计 II》，尤其是讲述江苏建筑职业技术学院图书馆（以下简称江苏职技图书馆）设计的《设计的逻辑》一文，会发现您提出的理性主义，更多是指向一种普遍的思想方法。这也是我尤其感兴趣的地方。因为我觉得中国建筑学中的理性主义还不够。之前对您作品的讨论，大多聚焦在"本土"上。我感觉，"本土"作为一种设计立场当然很重要。但这个立场是第一步，后面如何产生出高品质的设计，恰恰需要理性主义的方法。比如说在您工作室里，大部分建筑师都是认同"本土"理念的。但他们的设计水平肯定会有差别，跟您的项目把控能力也会有差距。所以我觉得理性主义是特别重要的一件事儿。

崔愷：你说得很对。实际上在我们工作室，并不是所有人都能够充分掌握本土设计或本土理性的原则。每次一个项目开始时，大家一块看现场回来以后，我会让大家都勾勾草图谈谈想法。很多时候，他们还是有一点飘忽不定，没有找到核心问题，或者说对关键问题不敏感，所以一般会去做各种可能性的覆盖式比较，效率比较低，要花许多时间去试错。我这两年提出来"精准设计"，项目分析一定是一刀切在关键地方，是针对这个项目提出的问题，进行理性分析，让这个项目所有的设计人都能了解我们要解决的问题和技术路线。这样不会跑偏，而方法是可以比较和尝试的，很快得到无论我们自己还是甲方都比较能理解和认同的设计方案。当然更大的问题是展开工程设计以后的水平，控制和引导——如何进行判断，如何跟结构工程师、设备工程师、工业设计师讨论问题，实际上都需要建立一个理性的体系来引导和推动。我一直认为创作是贯彻设计始终的，对于施工图亦或现场问题也需理性地作出判断。

范路：您提到江苏职技图书馆的设计，起源于3个方面：一是从形而上学层面追问建筑本质——"什么是大学生喜爱的图书馆"；二是从体验和感受层面提炼场地的空间特质；三是从工程、经济层面理

性把握建筑形式。那么在您头脑中，这3方面是如何聚到一起的？

崔愷： 这个项目是季翔老师介绍来的。季翔在中国建筑学会里面活动比较积极，也是江苏建筑职业技术学院的副院长。他跟我说有这么一个图书馆项目，3万m²左右，愿意不愿意做。实际上，之前已经有人做过一个方案，是那种相对比较封闭的形态。学校希望请我重新做。

到了现场之后，对场地的观察给了我很大的启发。整个场地在一个山坡上，在校门右手边有座小山，山上有几栋宿舍，体量不小，左手近处是体育设施，远处有个小水库，视野比较开阔。一般说来，对着校门口的建筑会是对称的。但这个场地是歪的，在斜坡上，挺难做的。前一个方案是先把场地抬平，再在上面做框架体系。但我觉得，可以尝试顺势做个不对称的建筑，它不需要形成刻板的轴线，而应该体现斜坡场地的地形特点。场地边上的小水塘和几棵大树，也让我有了某种感悟。原本没有图书馆时，大家可以在树下看书，有点像孔子在树下给弟子讲课的情景。所以我想，在场地中这个房子应该像树一样撑起来，像树一样展开，上面大，底下小，尽量减少对场地的干扰。树向四外伸展，方向感不强，形态有机而丰富，这也就回避了立面的方向和轴线对称的要求。

还有一点，虽然它对着学校大门，但这是新开的校门，去主校区还要往里走。我不希望因为盖了这个房子，大家都得绕着它去教学区和宿舍区。于是，我们就利用支柱体系把建筑架空，让人们可以穿过图书馆，走到后面的校园，也可以从教学楼穿过图书馆走到校门。要让你经过这个建筑，而不必绕开这个建筑，建筑底层可以穿越，成为校园环境空间的一部分。

此外，我们去国外也看到了当代图书馆功能的转变。现在的图书馆也常常被称为"学习中心"，不是封闭的，而是开放性的校园空间。大家来了并非只是看书，还可以举办各种学术活动和展览。我希望这座图书馆也有一定的开放性，于是结合场地特点把建筑架起来，把报告厅、咖啡厅和小书店放在底层。这样从报告厅出来，就能走到水边上欣赏景观，形成交流的场所。我们把书库和设备机房嵌在右侧的山坡里，变成半覆土的建筑，比较自然地利用了地形。建筑一层基本上是校园的共享开放空间，二层以上才是图书馆真正的管理范围，设置了门厅、存包处和中央服务台，然后是图书馆的阅览区。把想法聚焦在树的意象，解决了两方面的问题：一个是场地的关系，一个是图书馆作为学校的学习中心。

从树的意象出发，我们也考虑到绿色建筑设计。除了前面谈的底层开放空间外，在阅览室，考虑到学生喜欢靠近窗子看风景，就在窗口设置了花槽。二层以上有一些挑台，我们在挑台上也做了绿化花槽，并以挑台作为下层的遮阳构件。由于图书馆的正面是朝西的，西晒问题比较大。所以我们设置了一些格栅，让阳光不要直接照到阅览桌上。这些都是设计上的应对策略。今天看来尽管比较初步，还是落实了一部分绿色建筑和人文关怀理念。但是比较可惜的是，窗口和挑台屋顶绿化都没有做——甲方从一开始就很珍惜这个建筑，觉得浇花就把建筑弄脏了，养护成本又高，所以一直不肯做。我每次都跟他们讲，我对这个建筑的美学要求，是绿色建筑的自然美景，把绿化呈现出来，也是建筑表达的一部分。当然由于悬挑平台多，造成建筑体形系数大，不利于节能。徐州冬天气温低，悬挑会让楼板底面比较冷，脚底下不舒服。所以我们设计了地板供暖和窗口的风机盘管，希望冬天学生在窗户边上也很舒适。后来我们观察，冬天窗户边上的使用人数确实比较少，学生大多聚集在中部，这说明温度可能还是有些差别，没有真正解决这个问题。

范路： 所以说您的理性主义更侧重一种整合美学意境、生活体验和工程技术的思想方法，而较少关注样式和风格？

崔愷： 我对那种徒有其表的表皮设计是比较反感的。我较少做商业性项目，主要也是不喜欢那种为了商业而故意吸引眼球的策略，更倾向理性的设计方式。我觉得中国的民间传统是强调节俭的，因为我们也不是一个可以很铺张的国家。但为什么今天我们的城市这么铺张，我们对建筑的要求这么铺张？这实际上背离了中国的传统智慧。我们通常把中国传统文化挂在嘴上，但其实并没有真正用心去做，有很多地方需要回归，回到中国自己的传统价值观。但在设计手段上，却需要回到现代主义注重本体设计的方法。我记得关肇邺先生很早就说过，中国应该补上现代主义这一课。可我们改革开放初期就碰到了后现代主义、解构主义、晚期现代主义等各种潮流，后来商业化的时尚性建筑流派也很快传进来了，让我们眼花缭乱，尤其是国家大剧院国际设计竞赛时，中国建筑师一下看到40多个方案，发现"原来还可以这么做设计"，于是迅速地被各种潮流牵引，形成了不同的追风现象，在忙乱和浮躁中，忘了应该补上现代主义这一堂课。

直到2017年，60岁的我才抽空儿去美国短期访学。我有一个感触，就是美国近年的新项目并不是很吸引我。例如在旧金山市中心刚盖完的一个交通枢纽，是当地很有影响力的公共项目，但到现场一看挺失望，就是外头包了一层表皮，里面还是那些基本的交通设施。可我每次到旧金山市中心，都要去波特曼（John Portman）在市政厅旁设计的一个城市综合体坐一坐，从凯悦饭店通过连桥到达有好几层高的屋顶花园，人气旺，设计大气，感觉非常好。我也去北欧看过不少新建筑，但还是觉得阿尔托的东西好。所以我认为建筑设计作为空间艺术，似乎并不是一直在进步的，甚至有时候技术发展了，带来了一些新的可能性，反而会干扰建筑本身真正的艺术价值。我今天到了这个年纪，把建筑想清楚了，就知道什么是对人真正有吸引力、有价值的东西，而不是看到一种新东西就赶紧用进来，让大家觉得很好玩儿，更多的应是通过理性的设计，找到那种长久的价值。

范路： 通过理性主义，您能够提炼出江苏职技图书馆项目的本质问题。但理性主义的另一个重要作用，是充分贯彻设计构思，保证项目品质，让具体的设计手法不断变化但又保持逻辑一致。比如说，看您的设计草图，感觉建筑是像树一样长在草坡上，您希望学生能在树底下阅读，能在上部更多地欣赏周围的自然景观。所以建筑被架了起来，在矩形柱网体系下强化水平分层。同时，建筑每层向四面出挑并上下错开，这就创造出更多的欣赏四周绿地的阅读空间。而引入的斜撑结构配合变化丰富的外轮廓，又强化了最初建筑如树木般的设计意象。再看室内核心部分，围绕中庭四周，每层也有错动的小挑台。当然也有斜撑构件和直跑楼梯斜向元素。我觉得这是内部对外部的呼应，让建筑意象从外部传递到内部，从而带来连续一致的空间体验。

崔愷： 对，我的设计策略就是这样。在一个项目中，当我抓住了解决问题的设计方法时，就尽量多地用同样的手法来解决问题，包括内外的问题、总体和局部的问题。如果说在大策略上我有点浪漫主义的感觉，但真正贯彻设计的时候，我希望是理性主义的，要不断提炼建筑语言。在这个项目中就是这样。一开始，中庭并没有设计挑台，考虑到消防的需求还要做防火卷帘。但后来我觉得，中庭一定要有互动性，人看人，在安静的氛围中还有活力，所以在后期设计中就增加了一些内部挑台。

最开始我画设计草图时，因为有树的意象，想到了树和建筑之间的关系，所以很自然出现了斜撑构件。但是当时向业主汇报的时候，我记得学校领导有点担心，他一直质疑为什么会出现斜撑。他觉得这个建筑应该庄重大气，最好都是直的，不要斜向造型。他甚至觉得斜对他的地位是某种不好的隐喻。后来我就一直说，斜撑有利于节省造价，本来两根柱子都落地，就都得做基础。但我把它们合并起来，这个房子的基础就可以少做一些，也不影响上面整个空间的舒展，同时也介绍了"树"的意象，后来学校领导就接受了。但是我想避免让这个房子从外头看有一种表现主义的感觉，所以从外到内也不断重复外部的语言，当然也是和结构工程师配合的结果，在理性分析的前提下追求空间的丰富性和语言的单纯性。在工程当中，我们往往是要有这种策略的，就是你不能一开始就把问题复杂化、随意化，让业主理解不了。要让业主感觉到有一个基本逻辑，当别人问起这个事儿，他也能够进行解释，所以设计的理性要导向解读的理性，这样也容易达成共识。

范路： 在结构体系中，斜撑可看作是变奏的亮点。它减少了落地柱子的总数，并象征了树干形态和建筑所在地徐州的汉代木作建筑文化。而从工程技术角度来看，是否在这个项目中使用 BIM 系统，有助于斜撑的实现？

崔愷： 因为这个建筑的悬挑不是单向的，是向四面发散的，那么角部的斜撑就是双向倾斜的。如果柱子是方的，那斜撑和柱子角部的交接面就不容易控制好。所以我们跟结构工程师商量，把这些柱子做成了八边形，让它们有一个面专门朝向斜撑。但是把这些结构交接关系建立起来以后，就会发现用简单的二维绘图方法不太好控制，所以我们用了 BIM 设计。这个项目不是很复杂，设计费也不高，甲方也没有要求用BIM，但我们自己率先用了。这是我们本土中心第一次用 BIM 来建模，后来各专业团队也跟上来用BIM绘图，这不仅有助于落实建筑的空间结构关系，也能提升设计和施工的质量。虽然现在图书馆越来越自由，但我认为图书馆是有自身逻辑和基本模数的，它的开间跟书架和其他家具的摆放是有关的，方格框架是比较适用的空间体系。

另外在技术层面上，我们希望这是一个清水混凝土的项目。学校一直说他们没多少钱，所以我们想干脆不做吊顶，提高混凝土的质量，直接暴露出来。用BIM最开始是为了结构设计，后来还要满足穿梁等设备要求，所以就把机电设备做到了BIM系统当中。其实最开始给学校汇报时，他们对不做吊顶是有点担心的。但我说，施工上即便有一些不完美也没关系，这是一个培养施工管理人员的学校，大部分学生毕业后是去施工单位的，所以图书馆也是一个学习建筑知识的现场，让学生能够学习如何综合性地控制建筑质量问题。后来校方就同意了。实际上，因为做了 BIM，最后完成的效果是比较好的，结构施工的准确性很高，清水混凝土也实现了比较高的质量，大部分混凝土也没有特别修补。因为设计上失误较少，施工也比较顺利，我们去工地解决问题的次数也不是很多。总体来讲，虽然浇混凝土的施工时间比一般建筑稍微长一点，但后面只要拆了模板，基本上就算做完了。

范路： 在我看来，徐州项目中的局部斜撑处理是很适应中国国情的。因为在造价等限制条件下，它的主体部分——阅览室和书架区域，采用了正常方格网这一合理化的结构体系。而在内外一些关键部位，您用一些斜撑拓展了方格网体系，也增加了建筑形象和美学意境的表达。但是我们看到，库哈斯也做过不少图书馆项目。例如在著名的西雅图公共图书馆中，他大量采用倾斜的柱子，产生了对形式空间和建筑技术更为夸张的表达。那么您是如何理解他的作品？

崔愷： 西雅图公共图书馆我去过两次，应该说我很喜欢这个建筑，它在库哈斯的作品当中也是很精彩的一个。我注意到，以库哈斯为代表的一帮荷兰建筑师对于技术的表达是挺"酷"的。这种"酷"的感觉，来自有意识地呈现某种建筑空间结构的复杂性，而这种随意性是经过精心设计的。我对这种比较生猛的新美学还是蛮认可的。大家看到路易·康的图书馆是经典性的，所有的东西都是经过某种深思熟虑后得到的结果，做得非常严谨，严丝合缝。而库哈斯的建构设计有某种夸张呈现出来的随意感。很早以前我看过他在鹿特丹的美术馆，其中一根钢梁没有在建筑外墙界面停住，而是貌似随意地挑出来。他是在刻意表现某种建造逻辑的做法，这种感觉在工业建筑中经常会有，而在民用建筑中比较少见。包括中央电视台新大楼竣工时，我也进去看过，它的空间中有一些斜柱。本来是消极的东西，一般要是放在我们这儿，肯定是一通包裹，藏起来。但库哈斯全都露出来了，效果也很好。其实在建筑美学方面，中国传统建筑是有某种粗放性的，区别于日本建筑的周到和精细。所以我觉得，库哈斯建筑中自在的气质，还是值得中国建筑师学一学的。这是对建筑复杂性的综合呈现：首先是要宽容，其次把它纳入到建筑表现中，最后能变成你的建筑的气质，就是挺好的一件事情。

范路： 您刚才提到路易·康。其实在江苏职技图书馆项目中，我能感受到许多路易·康式精心的细部处理。当然，这是局部与整体逻辑一致的话题，也是理性主义的重要体现。比如在外立面上，对于出挑楼板端面的二次划分是等比例的，如同哥特束柱的处理方式，让楼板看上去更轻盈。而阅览空间采用玻璃加遮阳的设计，是在弱化建筑的体量感，强调楼板的水平性。再看室内，中庭的栏杆扶手处理成整条木板的形式，且后退以突显混凝土结构。而照明灯带安置在结构梁底面，是对井字梁形式的重复。实际上，这些处理都是在突出建筑的结构骨骼，强化设计草图中建筑如树木的意向。

崔愷： 在这个项目中，我们在内部界面处理和色彩设计上确实受到路易·康的影响。而这次整个结构设计，则是得益于我们和结构专业共同使用了BIM设计。否则的话，建筑师画完图后，结构工程师会按他的逻辑深化，大梁、主梁、次梁等结构构件会有不同的尺寸。在这个项目中，我们要求整个结构界面交圈，高度统一，这是很重要的。不仅结构界面统一，任何跟它无关的东西都要有所区别，我们不希望有别的东西把结构模糊掉。所以我们在室内设计中特别强调结构的体系性，让所有填充的东西都有明确的独立性。比如说空调管线，我们在有些地方会用格栅来遮掩杂乱的软管。还有灯的处理——我们在早先的一个图书馆项目中想把所有的灯都挂起来，形成一个独立的界面，但实际效果并不好，因为吊筋很难调直。所以这次我们就全部都做成吸顶式的，顺着梁做，宽度都是精确控制的，也算是一个比较成功的处理。

范路： 您还提到室内和景观设计方面感觉有点遗憾，能简单说说吗？

崔愷： 室内设计大部分是我们团队做的。实际上东西都很简单，但如果做得好就能跟建筑形成一体化的效果。我们设计了大厅、阅览室和咖啡厅，主要部分是控制住了，办公室就是他们自己做的。这个项目拖的时间比较长，后期换了几任领导，所以有时不太好协调，留下一些遗憾。我们在国外看到大师的作品，很多都是一体化设计的，非常重要。像阿尔托自宅和他工作室里的家具都是他设计的，做得很到位。芬兰的大学还特别强调，建筑当中放的家具都是大师设计的，据说都很贵。他们有这个传统，强调建筑中家具和工业设计的一体性。我们国内这方面还不太行，经常是建筑精心设计了，但到后期家具采购不是基建处管，而是由物业管理处负责。像在北京外国语大学的项目中我就多次遇到这种情况。所有施工完了以后，一摆家具就全跑题了。当然，我们也已经有了一些成功的案例，经过我们的游说，越来越多的甲方会同意由我们来选择家具。而景观设计上一个很大的遗憾是，旁边的那个小水塘没有保住。他们总说会有渗漏的问题，我觉得就是要那种水时有时无的湿地效果，不怕渗，而要那种"野味"。还有就是屋顶和窗前绿化一直不肯做，实在不能理解，也是这个作品的一大遗憾。

范路： 除了江苏职技图书馆，在您的《本土设计Ⅱ》中还有北京外国语大学图书馆和南京艺术学院图书馆这两个高校图书馆项目。这3个建筑的场地环境和内容要求都不一样，但它们的设计策略会有相似的地方，也会呈现出某种共同的内在气质。这是由于它们共享一种理性主义的设计方法，或是来自当代中国校园建筑在类型上的共性吗？

崔愷： 确实，我的基本态度是，校园建筑没有太多的商业压力，会比较理性。我们每次见到校领导都会去跟他们进行沟通，大家也基本认可这一点。因为造价的控制，建筑标准也比较一致，这就提供了一个基本的工作条件。当然还有学校的气质。总体来讲，大家对校园建筑气质的认识还是比较趋于一致，就是要典雅，要讲究细节，不能太夸张。这也让校园建筑的创作趋于某种理性。另一个我觉得很重要的前提是，我们比较熟悉校园建筑中的使用行为。使用者都是学生，不用考虑社会上很多的不确定性。所以在处理空间界面、楼梯甚至家具的时候，面对的是熟悉的人，他们能够很好地使用并爱护建筑。

还有，我到国外去看的多了，就越来越觉得校园建筑的寿命和长远价值十分重要。例如，我在北外改造过一个1970年代的建筑。现在看来，那个旧建筑确实不太讲究，因为那个时候确实没多少钱。那么到了今天，我希望这一次改造能使这个建筑存在得更长久些，再过50年回来看还能这样。对此学校也很认可。大家都觉得如果每三五年就要装修一次，那种短期行为是不会为校园留下真正的文化底蕴的。因为造价的控制，又有长久价值的追求，所以校园建筑设计一定不要太复杂，要简单可行，容易控制质量。有一些设计得很炫的建筑，虽然也能控制质量，但要花很大的代价，甚至要专门定制很多东西，这不太是校园建筑所需要的。所以这几年我碰到校园建筑，就知道大概该怎么做。最近，我们在东北大学新校区当中的图书馆也做得挺好。大家说这个房子猛一看不是很起眼，但经过50年、100年应该还能是这样。建筑外面是砌砖跟混凝土的结合，里面也把砖砌进去了。虽然校方一开始觉得造价有点高，但做完以后他们觉得这个房子能放得住。所以我觉得，校园建筑在中国当代建筑文化传承上可能是一个很重要的类型。

范路：《本土设计Ⅱ》是您在2016年对自己工作的总结。那么沿着本土设计或本土理性的方向，您设想自己今后的工作重点会是什么？

崔愷： 2008年《本土设计》的封面颜色是灰色的，拍的是首都博物馆的外墙石材，是对本土文化的某种呈现。2016年《本土设计Ⅱ》的封面我们用了黄色，是想表达建筑跟自然、跟土地的关系。我想我下一本书的封面可能会用绿色，因为我最近做的一些项目和课题，都与绿色环保节能相关。在做课题的过程中，我越来越觉得，其中建筑创作的空间很大。这不像以前大家认为，绿色建筑就是工程师的工作，实际上这也是建筑师的工作。而且我觉得，建筑师有时候片面地谈文化，会带来一种新的装饰主义，比如我们现在的一些礼仪建筑就是这样。习主席提出来"中国智慧"，实际上，我觉得能被所有人接受的"中国智慧"不是那些表面形式，因为别人显然不会接受你的形式。更重要的智慧，是来自于大家对环境和自然的关注，而这又回到我们老祖先了。所以每次讲绿色建筑，我都会放一张中国的古画。在山水环境中，人很小，房子很通透，人在自然当中生活。我觉得现在国内一些建筑有点儿走偏了，越来越喜欢很炫的东西。反过来，对于如何节约能耗、如何节地节材、如何让建筑响应气候特点等问题还缺乏关注。所以我目前的研究和设计方向，是想做理性主义的绿色建筑。它是人类共同的一种智慧。大家经常说"地域特色"，地域特色到底是什么特色？从根本上说，它是一种面对自然的态度，对生存状态的真实呈现，这就是地域特色。历史上所有的城市和建筑的地域性差别都由此而来，不是什么表面的装饰。地域特色又可以根据不同的技术条件和生活状态进行不断更新。我觉得中国建筑应该走到这条路上来，以面向未来的可持续设计去传承我们的传统文化，这种传承也就是可持续的。

——原载于《建筑学报》2019年第5期

大地生长——崔愷的敦煌"本土设计"建筑实践
GROWING ON THE LAND: CUI KAI'S "LAND-BASED RATIONALISM" ARCHITECTURAL PRACTICE IN DUNHUANG

支文军 郭小溪

摘要： 自2003年敦煌市博物馆的设计开始，崔愷及其团队历经近20年在敦煌设计了4个重要的公共建筑，始终立足敦煌本土且未曾中断。这一系列实践所生长的敦煌大地为"本土设计"的发展提供了根植的土壤，所产生的转变构成了观察"本土设计"走向成熟的内在线索。文章试图揭示崔愷的敦煌"本土设计"建筑实践与"人地"之间的紧密关联，并将其置于崔愷设计历程的三个不同时期中进行思考，勾勒出"本土设计"理念从萌芽、发展到成熟的生长路径。

关键词： 大地；本土设计；敦煌；历史文化；自然；理性主义

自1984年从天津大学毕业后开始执业至今，崔愷的建筑实践已持续36年。《时代建筑》曾于"承上启下:50年代生中国建筑师""建筑'新三届'"两期主题中对其作出了详细的访谈与报道，作为"文革"后入学的第一批优秀建筑师代表，他的建筑实践经历同中国城市与建筑快速发展的四十年相重叠，项目量庞大且涵盖了文化建筑、校园建筑、历史建筑更新、乡村建设等诸多类型，并建构出了自身成熟的"本土设计"理念。观察崔愷的设计历程，从某种意义上来说，也是一种对于当代中国建筑发展的反思。

笔者曾于2013年至甘肃参观敦煌市博物馆，其后于2020年7月再赴敦煌，探访了崔愷及其团队历经近20年在敦煌设计的4个重要公共建筑，这种体验经由时间跨度所强化，受西部辽阔的空间尺度所加深。茫茫大漠与无垠戈壁，半山佛窟与千载古道印刻着悠久的历史，广袤的土地上党河缓缓流淌、光影昼夜变幻，不经意间轻易震撼着观者的内心。崔愷的敦煌实践在这片壮阔的大地上生长，最终凝为了敦煌本土文化的重要组成部分。

1. 大地："本土设计"之源

"建筑在大地上隆起，又终要消失回归于大地之中；建筑扎根本土，在本土的滋养中破土而出，茁壮成长。"
——崔愷

"本土设计"（Land-based Rationalism）是崔愷提出的强调立足土地，以自然和人文环境资源的沃土为本，以"理性主义"为设计原则的实践策略。他将这一理念的思想核心归于"土"，即人类所赖以生存的"大地"。"大地"不仅是"土壤"，是人类脚下牢靠实在、充满生机的基地与根柢，也是建筑师精神世界的庇护者，是使"世界"张开自身的物性基础。"本土设计"的思考试图通过建筑作品阐明"大地"并揭示其本质，回应着在现代主义建筑作为话语主体的视野里，往往被隐匿着的"大地"应如何显现，而建筑又该如何归萌于"大地"的庇护的问题。

"本土设计"在崔愷2016年出版的第三本作品集《本土设计Ⅱ》中有着清晰的图示呈现：书中一棵枝繁叶茂的大树，扎根于由社会政治经济资源、地域文化资源、科学技术资源、土地环境资源、气候资源、工艺材料资源所构成的大地，这些丰富资源凝聚成了"本土设计"所根植的土壤。经由"立足本土的理性主义"的躯干吸收转化，进而生发出地域建筑、社会建筑、新乡土建筑、生态建筑、地景建筑等枝条，最终构成了一个繁茂的有机体——一棵衍生于大地的"本土设计之树"。总体来说，"大地"是"本土设计之树"的立足点，是进行理性分析并作出设计决策与判断的支撑，最终孕育出由多元建筑语汇与开放的表现策略所建构的完整系统。

崔愷的"本土"，特指于中国的"大地"本土，是自然和文化环境资源的综合。不同于"地域主义"绝对的历史主义式的过往怀念，"本土设计"反对保守倒退，始终立足于本土"大地"展望未来。也异于"批判的地域主义"对"全球化"的抵抗，"本土设计"具有"现实理性"，基于"现代世界"这一既成事实之上。"本土设计"理念同我国传统"大地"观一脉相承，观察西方风景画中，建筑往往处于开阔的远景处成为整个构图的焦点，追求壮观雄浑的效果。而反观中国山水画，人在自然山水中生活，建筑往往通透轻盈地融入大地，成为大地中的一部分。这种传统文化中人与"大地"的自然和谐关系是崔愷"本土设计"理念的根源，衍生出综合中国广阔大地上的自然生态、地理环境、文化历史因素，并对不同地区呈现出的明显地域差异作出回应的本土"大地"观。

2. 大地生长的敦煌实践

敦煌位于甘肃省西北部，河西走廊的西端，大地辽阔壮美，是古丝绸之路上的重镇。东汉应邵曾作《汉书》注："敦，大也。煌，盛也",勾勒出敦煌广开西域的繁盛景象。由于四周为沙漠、戈壁，风沙时而漫天，敦煌的环境气候非常特殊，是我国最为干旱的地区之一。同时，敦煌的文化艺术璀璨丰富，在中国建筑史上也具有着重要的价值，莫高窟中的建筑、雕塑、壁画以及敦煌古建筑实物和遗址为我国古代建筑研究提供了重要资料，中国营造学社就曾以第61窟的"五台山图"作为指引，首次寻找到了国内尚存的唐代木构建筑——佛光寺东大殿。

自2003年敦煌市博物馆的设计开始，崔愷在敦煌完成了长期且持续的设计实践，崔愷的4个公共建筑——敦煌市博物馆（2003-2010年）、莫高窟数字展示中心（2008-2014年）、敦煌市公共文化综合服务中心（2013-2018年）、雅丹地质公园游客中心（2014-2017年）就生长于这片广袤的土地之上。其中，文化服务中心与敦煌市博物馆紧邻，位于敦煌市区南部，而莫高窟数字展示中心与雅丹地质公园游客中心分别处于莫高窟与雅丹地质公园近旁。前两者处于城市环境，后两者则与历史古迹、自然地貌景观相邻相依。

敦煌市博物馆的设计与建造跨越了八年的时间跨度，初建时还处于城市的边缘，待建成后已被新建筑所包围。博物馆呈现出外部封闭、内向敞开的院落式布局，观者可以自大地起始，随展厅内螺旋上升的路径漫步至屋顶平台，眺望城市与鸣沙山景观。而其后的莫高窟数字展示中心则处于一片视野开阔的场地之中，距离莫高窟约15km，建筑南侧为平坦的戈壁滩，远远望去，可以窥见沙峰重峦的三危山与鸣沙山。建筑形体沿场地南侧长向一字舒展，如大地流沙般起伏，缓缓从大地中升起。相较敦煌市博物馆所呈现出的外在封闭的内向化空间，展示中心的内部空间则更为开放，同建筑流动的形态保持着逻辑上的一致性。

而敦煌市公共文化综合服务中心的设计开始于数字展示中心建成前一年，建筑位于博物馆南侧，同博物馆及数字展示中心较为封闭的外部形体不同，建筑呈现出开放的形体状态，所采用的方形母题在敦煌市博物馆中便已经有所显现。文化中心通过不同尺度原型的叠加、院落的空间序列组合，以及外部的积极空间呈现出了一个具有公共形态的"文化聚落"。雅丹地质公园游客中心与文化中心的设计建设几乎同期。建筑位于敦煌以西约100km外，距离雅丹地貌的核心景区还稍远，场地身处戈壁腹地，周边草木不生。建筑形体顺遂大地的重力，高度上低于周围崖体，谦卑俯卧于戈壁滩的辽阔大地之上。圆盘状的建筑体量被切割成为若干条，遥遥指向雅丹地貌景区，与经过千百万年风化而成的舰阵般的崖体相呼应。

这4个公共建筑同敦煌大地之间的关系紧密，它们或对传统建筑形制、城市地域特征进行提取并作出抽象转译，或对历史要素、本土语汇进行吸收并作出理性表达。它们呈现出因场所特征的区别以及随之采取的不同策略而产生的明显差异，但对敦煌本土历史文化与地域特征的挖掘与回应则构成了其设计所隐含的逻辑一致性，显现出一种内在的"本土理性"。

由于种种原因，雅丹游客中心至今还未能全部开放，但博物馆、数字展示中心和文化中心这三组建筑已经使用多年，其间经历了各种文化交流活动的举办，接待了大量游客的到访，建筑的社会价值得到了良好体现。数字展示中心借助数字技术与多媒体展示的方式，有效缓解了莫高窟的开放同保护之间的矛盾，建筑师团队还对项目中绿色技术的实施作出了科学的使用后评估，建筑在节能效果上达到了预期效果。这些建筑的社会意义的实现弥补了敦煌本地的施工技术有限、项目资金不足等限制所带来的细节上的遗憾，而建筑文化价值的实现使建筑成为敦煌文化的自身构成部分。同时，这一系列实践的设计时间前后相连未曾中断，且至今仍有设计在敦煌持续进行，这种时间线索与其"本土设计"策略的发展、成熟过程有着隐性的重合。理解崔愷的作品，需要以连贯的视角去理解他"本土设计"理念与本土大地的关联，从敦煌的连续性实践切入，不失为一种有效的观察方法。

3. 崔愷建筑实践的三个时期

综合崔愷长期且持续的建筑创作历程，我们可以将其宏观地划分为三个时期。第一个时期为1984年至20世纪90年代初，以阿房宫凯悦酒店（1986-1991年）、北京丰泽园饭店（1991-1994年）为代表，前者受南京金陵饭店影响，展现出抽象的大屋顶形式，而后者则受到贝聿铭所作北京香山饭店的影响，表达出对于城市环境的思考。在这一时期，崔愷通常采取以传统符号创作外在形式的设计方式，但民族符号的大量运用在一定程度上掩盖了对于建筑物本体的体量表达。第二个时期以北京外研社办公楼（1993-1997年）为节点，建筑由几何体量切割而成并呈现出清晰的形体关系，以空间逻辑的强化彰显出建筑自身的文化属性，试图营造出城市空间与建筑空间的渗透融合。外研社办公楼的设计奠定了崔愷设计观中的"理性主义"基础，其在这一时期的设计方法从强调外在形式向关注空间构成转向。

前两个时期可以称作崔愷设计实践的早期，他本人对于"形式"

的天分与求学时便展现的"现代主义"倾向、天大传统教学训练中对绘图表达的注重、硕士导师彭一刚先生的引导等诸多因素,奠定了其早期实践的外在形式语汇与内在设计理性的重要基础。他的设计手法同20世纪80年代西方"后现代主义"建筑理论与作品传入后"风格"问题成为建筑界焦点的时代背景相呼应,设计思考与当时学界对传统文化继承和建筑"神似"还是"形似"的话题探讨相关联。改革开放初期的深圳实践使其积累起大量的实际项目经验,80年代末同中国香港现代建筑的接触体验以及1995年后的美欧考察逐步扩展了他对城市历史文脉、现代城市生活的感悟与认知。2002年,崔愷出版了著述《工程报告》,以个人随笔为起始,将个人作品作为叙述主线,通过工程图纸的方式对个人早期的实践作品进行了"报告"式总结。在书中,崔愷还偏向于针对项目本身在工程完结的时候写一篇回顾性的思考总结,清晰的个人思想框架还未建构成型。

第三个时期的时间跨度为21世纪初至今,以北京德胜尚城(2002-2005年)为节点。同前两个阶段侧重于对"现代主义"设计手法与"后现代"符号表达的结合所不同,在这一时期,崔愷开始思考和探索自身的设计发展道路,"本土设计"思想逐渐成型。他的设计形式语言愈发抽象,设计手法更加灵活多变,开始更加强调建筑与人文、自然、环境的特定关系,并基于场地环境作出理性分析,逐渐走向了一条"立足本土的理性主义"之路。

4. 敦煌实践与"本土设计"理念的演进

观察崔愷设计历程的第三个时期,可划分成三个不同阶段:"本土设计"的萌芽、发展与成熟。敦煌的本土实践均处于这一时期,四个公共建筑的设计历程与"本土设计"理念的产生与成型相回应,并且完整贯穿了"本土设计"观所发生的三个阶段。敦煌市博物馆的设计处于第一阶段,见证了崔愷对于场所历史文化的关注转变。莫高窟数字展示中心是第二阶段的代表,也是崔愷关注将自然融入设计的重要突破的标志。而敦煌公共文化服务中心与同期的雅丹地质公园游客中心,则呈现出在"理性主义"方法支撑下更具综合性与灵活性的设计策略表达。

4.1 "本土设计"理念的萌芽期(21世纪初-2008年)

这一时期以北京德胜尚城为节点,直至2008年其第二本作品集《本土设计》中对于"本土设计"理念的正式提出。在经济全球化的背景下,尤其是2000年中国加入WTO后,国内的设计院积极投身于蓬勃发展且竞争激烈的设计市场。2003年,崔愷所在的中国建筑设计院进行了专业化调整,并成立了三个名人工作室,崔愷工作室就是其中之一。这种内部的专业化主要是为了引导院内专业技术水平的提升,并试图实现建筑师的"个人身份价值"。这种调整提升了设计院面向市场的效率,随之而来的大量项目实践促使崔愷开始有意识地对个人实践进行思考总结,"本土设计"的理念在这一时期开始萌芽。德胜尚城就是在这种背景下设计完成的,建筑提取了城市历史街道、老北京四合院等原型,回应了拥有六百年历史的德胜门城楼。这一建筑是其第一次从建筑单体走向创造开放的城市社区,标志着崔愷在关注建筑本体的基础上,向更加关注建筑与环境的关系、场所历史文化的转变。

敦煌市博物馆

莫高窟数字展示中心

敦煌市博物馆的设计于崔愷工作室创办的同年开始进行，与德胜尚城的设计时期相近，建筑开启了崔愷在敦煌实践的序幕。博物馆位于敦煌市区南部，中心院落式布局表达出敦煌本土民居院落形式的抽象转译，回应了莫高窟壁画中敦煌民居的院落格局：明确的庭院中心，内向性回廊兼做外墙。自入口西域画廊起始，以时间线索串联起展示敦煌历史沿革的各个展厅，整体的回字形流线将观者引入博物馆水平向延展、垂直向上升的展陈空间之内，建筑在纵向空间上逐阶抬高，将空间汇聚至绘有飞天的顶部中央藻井，中央展厅呈现出对莫高窟内唯一未曾有发展断裂的覆斗顶形窟的抽象表达。博物馆通过建筑形体、空间构成与组合方式抽象出敦煌的"本土"元素，建筑肌理与材质的表征同本地居民的经验记忆相关联。建筑采用的回字形路径与中心性庭院的布局方式在其后的安阳殷墟博物馆（2005-2006年）中也有所体现。前者的体量与材质对汉代长城、烽燧遗址作出了遥远的回应，后者则以植被覆盖并隐匿于地表之下，最大限度地减少了对于遗址区域的干扰。在这一阶段，崔愷完成了一系列的历史博物馆、考古遗址博物馆，如首都博物馆（2001-2005年）、无锡鸿山遗址博物馆（2005-2008年）等，逐渐加深了其对于城市文化象征、地景、历史环境的重要价值的理解与感悟。同期的西藏拉萨火车站（2004-2006年）、苏州火车站（2006-2008年）设计也展现出建筑师对于城市文化内涵重要性的思考传达。

4.2 "本土设计"理念的发展期（2008-2014年）

第二个阶段以2008年崔愷"本土设计"理念的正式提出为节点。在2008年其出版的第二本作品集《本土设计》（Native Design）一书中，崔愷对之前的实践作出回顾并提取出"本土设计"的三层涵义："环境"即人类脚下的土地，"文化"即大地中隐含着的深刻的本土文化，以及"空间"这一建筑地域性与场所感的三重概念。伴随着实践量的增长，丰富的创作积累也促使崔愷不断转身思考这些作品中能概括和引导自己创作的核心，"本土设计"成为其之后实践的理论支撑。在这一阶段建筑师自信与创新精神得以增长，进而形成了在"理性主义"方法支撑下关于"本土设计"的理论框架，这种设计理性使其能够在具体的设计手法不断变化中保持逻辑的一致。展现出如北川文化中心（2009-2010年）、康巴艺术中心（2011-2014年）中对于少数民族文化的现代转译，既现代又具有当地文化特色。以及浙江大学紫金港校区（2008-2010年）、江苏建筑职业技术学院图书馆（2009-2014年）等对于校园建筑与自然、公共空间关系的成熟思考。

这一阶段是崔愷"本土设计"理念的发展期，敦煌莫高窟数字展示中心可以称为这一阶段的开端。建筑临近莫高窟，紧邻平坦的戈壁滩，远处为三危山与鸣沙山。建筑的形态如大漠流沙一般从大地中生长，以有机的自然形体统一起多样的公共空间。展示中心内出现了两种空间序列层次的并置，这种空间的变化源于对莫高窟的观察——"单一洞窟"与"组合洞窟"的并置。"单一洞窟"展示出空间的单向性，通过洞口直指沙漠的尽头，"组合洞窟"之间彼此串联，创造出灵活多变的空间体验。建筑内部以抽象的"虚体洞窟"对应着功能空间，呈现出外向的静态氛围。而"实体"的岩土空间则展示出内向的不断流动的状态，与自然连续的展示空间相对应。这种感受被赋予

敦煌市公共文化综合服务中心

雅丹地质公园游客中心

建筑的动线之中，强调出内部空间的流动性，外墙的随机洞窟立面构成也暗示出内部空间的灵活多变，建筑整体呈现出理性主义策略下"本土设计"的整体逻辑一致性。这种对于自然的回应在泰山桃花峪游客中心（2009-2010年）中也具有明确体现，后者位于通向泰山的道路旁，建筑形态以山形组构，通过混凝土材质摹拟石质肌理，回应了泰山的裸露岩体与自然地貌条件。在设计时间相近的昆山文化中心（2009-2012年）、中信金陵酒店（2010-2012年）项目中，均有意识地将自然有机融入到建筑设计之中。以莫高窟数字展示中心为代表的几个作品，展示出其在这一阶段自然的而非文化的背景被引入设计作品，实现了建筑师对自然隐喻的抽象情感表达的重要突破。

4.3 "本土设计"理念的成熟期（2014年至今）

这一阶段以2014年工作室正式更名为节点至今。崔愷工作室在2014年更名为"本土设计研究中心"，由三个设计室与一个研究室构成。研究中心以"本土设计"命名，以"立足理性、明确本土"为思想核心，形成了尊重特定环境中的历史文化资源，创作富有本土地域特征建筑的团队价值观。其后于2016年出版的《本土设计Ⅱ》（Land-based Rationalism Ⅱ）中，崔愷作出了自2008年后的实践作品梳理，呈现出对"本土设计"理念的阶段性总结，发生了从"Native"到"Land-based"的对于本土"大地"更加明确地强调，而作为内在方法支撑的"理性主义"思想也愈发明晰。相对于《本土设计》对设计作品的集中呈现，《本土设计Ⅱ》中将其实践进行了梳理归纳，分为五个主题：自然主题、历史遗产、地域文化、有机更新、城市生活，展现出"本土设计"发展期所作实践中已呈现出的多样性与开放性。值得一提的是，莫高窟数字展示中心是其中所呈现的第一个作品，是上一阶段崔愷"本土设计"理念的重要代表。同时，作品集中的思考与言论显现出其对"本土设计"更为成熟的思考与解读：立足于从项目所处环境中提取有特点的要素作为切入点，以当代的设计语言选取恰当的建筑语汇表达，实现一种融汇建筑、历史文化、自然环境、城市生活等多方面要素的因时因地的和谐结果。

可以说，这一阶段的崔愷达到了设计观的定性，但设计手法则更加多元开放，实现了"本土设计"思想的成熟化，敦煌公共文化服务中心是这种"本土设计"观的有效体现。文化中心位于敦煌市博物馆南侧，设计过程经历了从巨型的超尺度设计方案到统一模数控制下的小尺度分散体量的转变，最终呈现出更为开放的公共状态。建筑以院落为中心组织起空间序列，回应着在敦煌市博物馆中便已有所显现的方形母题。图书馆开大窗，档案馆开小窗，剧场以封闭盒体组构而成，在太阳辐射很强的沙漠气候中能够有效节能的同时，以大小方窗的组合方式化解了单一体块的生硬感。与敦煌市博物馆的封闭内向性布局相比较，文化中心呈现出的聚落形态产生了自然与人性化并存的秩序，实现了其"乡土聚落"感的营造。这种开放的聚落形态在北川文化中心、康巴艺术中心的多种功能汇聚、院落空间的有机组合形态中已有所体现。这种隐含的聚落形态与结构回应着人们与自然共享后所形成的文化价值观念，是崔愷对当代建筑公共性与传统民族文化的积极回应。而雅丹地质公园游客中心临近雅丹地貌自然景观，建筑物紧紧地拥抱大地，与巨大的雅丹地貌尺度相对应，同海口市市民游客中心（2017-2018年）、世界园艺博览会中国馆（2016-2019年）中对自然有机形态的摹拟有着一致的与自然相和谐的设计表达。这一时期的代表建筑有着如中车成都工业遗存改造（2018-2019年）、雄安设计中心改造（2018年）项目通过保留城市历史印记使建筑再生，西浜村昆曲学社（2014-2016年）、昆山锦溪祝家甸村砖窑改造工程（2014-2016年）的乡村营建对于地域特色的传承与文化回归的乡愁意识表达，能够看出，在"本土设计"理念的统一下，诸多设计观念同时出现并相互作用，产生出将环境观、文化观、城市观、乡土观、绿色观为一体的整体策略。

5　结语

崔愷的"本土设计"理念，是在中国当代建筑快速发展且没有太多历史经验可借鉴，而西方现代建筑思潮不断涌入的境况下，建筑师基于大量本土实践与思考而逐渐形成的理论体系。敦煌实践展示出崔愷在敦煌大地上近20年的连续性实践，勾勒出"本土设计"理念从萌芽、发展到成熟的生长路径。基于这一理念，崔愷对于敦煌的深厚历史文化、自然地域环境、建筑与城市的公共意义作出了有意识的积极回应，而敦煌实践也为"本土设计"理念提供了可扎根生长的大地土壤。在敦煌的四个公共建筑中，建筑师将个人特征隐匿，使本土地域文化特征得以显现，建筑最终于敦煌大地中自然生长。

在近40年的实践中，崔愷的设计思想经历了从以民族符号创作外在形式，到关注建筑空间构成，再到思考建筑同环境及历史文化之间的关系，直至对历史文化资源、自然地域资源、社会资源等作出综合性思考的成熟的"本土设计"理念的过程。并且随着实践的不断积累，崔愷对于建筑项目的态度也从单向性的兴趣向综合性的

	1984—20世纪90年代初	20世纪90年代初—21世纪初	21世纪初—2008	2008—2014	2014年至今
书籍出版			《工程报告》2002 Projects Report	《本土设计》2008 Native Design	《本土设计II》2016 Land-based Rationalism
执业经历 机构成立与扩张	建设部建筑设计院　建筑师 华森建筑与工程设计顾问有限公司（深圳） 华森建筑与工程设计顾问有限公司（香港） 建筑师	建设部建筑设计院 副总建筑师 天津大学 兼职教授	高级建筑师　主任建筑师 中国建筑设计研究院　副院长、总建筑师 中国建筑学会副理事长 国际建协副理事 国家工程设计大师 崔愷工作室成立　2003	国际建协国际竞赛委员会联席主任 中国工程院院士 清华大学建筑学院双聘教授	崔愷工作室更名为"本土设计研究中心" 2014 保加利亚国际建筑研究院院士
敦煌实践			敦煌市博物馆 （2003—2010）	莫高窟数字展示中心 （2008—2014）	敦煌市公共文化综合服务中心 （2013—2018） 雅丹地质公园游客中心 （2014—2017）
"本土设计"理念 演变的三个阶段			"本土设计"理念的萌芽期	"本土设计"理念的发展期	"本土设计"理念的成熟期
建筑实践的三个时期	1984—20世纪90年代初	20世纪90年代初—21世纪初	21世纪初—2008	2008—2014	2014年至今
阶段特征	多采用以传统符号创作外在形式的设计方式 如阿房宫凯悦酒店的抽象大屋顶形式、北京丰泽园饭店对于本土民居元素的回应 但民族符号的大量运用在一定程度上掩盖了对于建筑物本体的体量表达	从强调外在形式向关注空间构成转变 北京外研社办公楼的设计试图营造出城市空间与建筑空间的渗透融合，奠定了其设计观中的理性主义基础	北京德胜尚城标志着其在关注建筑本体的基础上向更加关注建筑与环境的关系、场所的深层历史文化转变 设计了如敦煌市博物馆、殷墟博物馆等一系列历史博物馆、考古遗址博物馆，对于城市文化象征、地景及历史环境的重要价值的理解感悟得以加深	形成了在理性主义方法支撑下"本土设计"的理论框架，这种设计理性使其在不同场地中产生具体设计手法的变化，但保持逻辑一致性 以莫高窟数字展示中心为代表的几个作品展示出在这一阶段中，自然而非文化的背景被引入设计中，取得对自然隐喻的抽象情感表达的重要突破	在这一阶段达到了设计观的定性，设计手法更加多元开放，实现了"本土设计"思想的成熟化 在"本土设计"理念的统一下，设计呈现为诸多设计观念同时出现并相互作用的结果，产生出将环境观、文化观、城市观、乡土观、绿色观融为一体的整体策略
各阶段代表项目	阿房宫凯悦酒店（1986—1991） 北京丰泽园饭店（1991—1994） 蛇口明华船员基地（1989—1992）	北京外研社办公楼（1993—1997） 外交部怀柔培训中心（1994—1995） 威海中信金融大厦（1996—1998） 北京现代城幼儿园（1999—2001） 北京现代城嘉屋公寓（1999—2002） 北京外研社二期印刷厂改建（1999） 清华科技园创新中心（2000—2002） 北京富凯大厦（2001—2002） 北京雅昌彩印大厦（2001—2004）	北京德胜尚城（2002—2005） 首都博物馆（2001—2005） 辽宁五女山高句丽遗址博物馆（2003—2008） 韩美林艺术馆（2003—2008） 大连软件园8、9号楼（2003—2004） 安阳殷墟博物馆（2005—2006） 拉萨火车站（2004—2006） 山东省广播电视中心（2004—2008） 凉山民族文化艺术中心（2005—2007） 山东理工大学图书馆（2005—2007） 重庆国泰艺术中心（2005—2013） 苏州火车站（2006—2013） 中间建筑艺术家工坊（2007—2009）	浙江大学紫金港校区农生组团（2008—2010） 鄂尔多斯市东胜体育场（2008—2011） 北川羌族自治县文化中心（2009—2010） 泰山桃花峪游客中心（2009—2010） 昆山市文化艺术中心（2009—2012） 德阳市奥林匹克后备人才学校（2009—2011） 欧美同学会改扩建（2009—2013） 江苏建筑职业技术学院图书馆（2009—2011） 鄂尔多斯市胜利饭店（2010—2012） 北工大综合教学楼区（2010—2012） 中国杭帮菜博物馆（2010—2012） 中信金陵酒店（2010—2014） 唐艺术中心（2011—2014） 西安大华1935（2011—2014）	昆山祝家甸村砖窑改造（2014—2016） 西浜村昆曲学社（2014—2016） 湖南永顺县老司城遗址博物馆（2014—2015） 荣成青少年活动中心（2016—） 世界园艺博览会中国馆（2016—2019） 海口市民游客中心（2017—2018） 昆山大戏院（2017） 中车成都工业遗产改造（2018—2019） 雄安设计中心改造（2017） 雄安市民服务中心企业临时办公区（2018）

思考转变。这不禁令人思考，从崔愷早期的阿房宫凯悦酒店"大屋顶"、外研社"中国红"到今天成熟的"本土设计"实践，对历史文化与地域环境的抽象转译逐渐成为重点而建筑师的个人痕迹渐渐隐退，在以团队合作的方式实现"本土设计"普适性的价值立场和包容度的同时，这种"匿名性"也相伴发生，在一定程度上引发了项目之间的"差异感"。如何在有效解决现实问题的同时，依旧能够保持本土地域特征的传承与建筑的内在逻辑一致性的持续表达，敦煌实践不失为一种有效经验，也相信"本土设计"理念在未来能够给予我们更多解答。

——原载于《时代建筑》2021年第1期

方向与方法
——源于本土设计实践的谈话

被访人：崔愷
采访整理：刘爱华
访谈时间：2022年10月18日
访谈地点：中国建筑设计研究院本土设计研究中心

刘爱华：您在本书中非常系统地总结了本土设计的一系列方法和策略，那么在具体项目中，面对复杂、多重的现实条件，您认为应如何灵活应用这些方法和策略？

崔愷：我们强调本土设计，实际上是强调建筑和外在条件之间的互动关系，这些互动关系通常是多层级、多方向的，它们对建筑的影响也呈现出不一样的程度。在具体项目设计的应用中，这些互动关系的复合性虽然很多，但可以通过筛选，在最开始找到主要矛盾，在过程中在坚持解决主要矛盾的大思路下，去解决次级矛盾，做出应有的判断，又不失对主要矛盾的坚持。

我一般会在项目开始的时候放松自己，不去刻意地找所谓的设想。当接到一个设计任务后，团队会去搜集相关资料，但我在这时不会把大量的参考案例装在脑子里，也并不特别着急去做出主观的判断。我认为特别重要的是保持在现场的新鲜感，通过在现场的行走、跟业主方的交谈、对环境的观察，带来很有启发性的感悟。感悟对设计来讲是特别重要的，在这个过程当中，会判断出主要的思考问题到底是什么，到底是环境特点带来的问题，还是主体功能所带来的问题？

现在对于大部分建筑尤其是文化类建筑，功能排布上基本没有解决不了的问题，但是在特定的环境下，如何借用环境优势、解决或避免环境劣势，去形成功能的合理排布，往往就要对功能和环境这两者之间的关系做出判断。

比如做博物馆设计，我会对内向和外向的环境空间作出判断，是开敞的还是封闭的，安静的还是有仪式感的。再比如做图书馆设计，基地的外在环境是观景的还是喧嚣的，在哪些高度或位置环境是混乱的，在哪些位置环境是安静的？这些问题我都会在脑子里反复提问。

因为我们了解人最喜欢在什么情况下阅读，也知道如何在这个场地上去找到这样的场所，所以有时会把阅览区放在高处，有时会沉在半地下，用这些选择来修正环境当中不好、嘈杂的一面；如果场地处于特别安静的环境中，要做的就是让阅览环境融入环境当中。这些判断都将成为设计中很有意思的表达。

在武汉大学城市设计学院的场地勘察过程当中，实际上当我走完了场地就知道应该怎么做了：场地一侧是山，另外一侧是非常密集的学生宿舍，那么主体教室功能空间应放在山侧，另一侧排布小的老师工作室，我当时还想到要怎么把这部分藏起来，跟宿舍之间形成视觉上的隔离；纵向来看，结合从山上下到场地本身的高差，思考怎么样把高差的感觉延伸到教室，而不仅仅用楼梯间去连接各层，让这个空间体就像一座山一样；放在"山"上的教室的想法又跟设计教学本身相关，形成既创造领域感、又创造共享的阶梯教学空间。类似这些我都会在现场做出判断，把这些想通以后特别兴奋，这一趟场地走完之后，方案想得很清楚。

设计的第一介入点是来自于现场，而不是来自于建筑师自己的某种凭空想象，或来自于哪个大师的启发。这跟很多建筑师的方法不一样，有的建筑师心中有一个理想空间，当把实现理想空间作为第一选择，把环境作为第二、三选择，实际上就相对比较被动，但也不能说不成功，有的建筑本身挺成功的，可是有时会觉得放在这个场地上，好像应有更好的办法。

当然也有的业主或领导并不是特别介意我们所发现的这些场地特征，他们觉得都可以改变，确实也有这样实施的。所以如果用自洽性的建构逻辑来做设计，有时候阻力也未必那么大，只要形态呈现得好，甲方也觉得挺好。但对我来讲，以土为本的设计方法是不一样的，所以我常常要说服甲方理解环境的重要性、建筑和场地协调的基本逻辑。实际上现在中央提出来一系列保护生态的"绿水青山"、城市更新中"下手要轻"、乡村振兴中要"留下乡愁"重要指示等，这些都要求项目建设要跟外在环境直接有关。本土设计的逻辑就是将满足与周边环境相和谐的要求作为第一要务，而且我也相信这样的和谐姿态并不会从根本上否定内在功能的合理性，所以这套设计逻辑也得到了越来越多的支持。

我在本书中《关于场地踏勘》文章里也专门讲到，看现场一定是带着问题，去之前了解任务，然后带着一定的问题到现场去看，这对我自己是很重要的。我也有些只做指导的项目自己并没有到现场，但是相比较听别人介绍场地，到现场是更好的选择，也正因为对现场判断的依赖，如果没有去过的话，我在讨论中听团队介绍现场环境时会问的问题比较多，甚至有些问题追问到他们根本没准备或者是视而不见的，这类情况我多次碰到，说明团队对现场的特征往往还不够敏感。

我还有一次经历，我和关飞建筑师到瑞士去给丁肇中先生汇报日照项目方案，出发前院里跟我说外交部要在斯特拉斯堡建一座领事馆，但对法国事务所做的方案不太满意，希望我去看看。我们在日内瓦办完事后开车到了斯特拉斯堡，基地时下着漫天大雪，当走到一条雨水沟和一片大树处，我问是不是已经到了场地边界，陪同的人说这只是中间，后边还有一半，我说那怎么在原方案中没有看到这片树林？虽然这片树不见得有多名贵，但是我觉得把它们保留下来对项目的布局规划是有意义的，至少不能先把它取消。第二天早上我见到法国事务所的老板时提到了这件事，他说问过这个地区的规划师，这些树可以不保留。我说对我们来说，对树的是否保留建筑师自己要做出主动的判断。这个项目我们后来没有参与，之所以今天提到这件事，是觉得我们常认为欧洲建筑师都很成熟理性，但实际上有些时候也很任性。有的建筑师对环境的应对就是满足于符合规划条件，但对我来讲，找到场地上哪怕是微小的差异性而使建筑变得很特别，这是本土设计努力的方向。法国人的方案算不算好建筑呢，我想建完可能也还可以，但它跟这个场地的遗存物没有任何关联。而我希望我们的设计来自这片土地上的差异性特点，对任何改变我都要经过深思熟虑，首先都要以敬畏的心态，要保留下这些差异性，这是一个立场问题，也是一个价值观的问题。

当然看现场仅仅是设计的开始，找到主要设计点之后，还要考虑怎样去解决其他层面的问题。在设计深化当中肯定会有些地方要与团队反复讨论，但因为我的习惯是找到主要设计点后就会把草图画出来，所以后面的深化都是在这个主要的空间构想之内去处理一系列的技术、功能、节能环保等问题。当然还要结合考虑业主的想法，比如面对业主提出来的商业模式或经营模式，要去思考他的重点在哪儿，怎样能沿着他的思路去把空间构成做得更加合理。这些是次一级的矛盾，不作为第一层级的矛盾来判断。我处理这些事的时候，在层级上分得比较清楚，而且轻易也不会动摇。不会出现我在现场想一个主意，然后跟业主要求完全不吻合，甚至出现了先天的矛盾，最后造成翻车。业主完全不接受我们设计的想法，这种情况在我这儿很少出现，当然这跟经验有关系，我知道对于某种类型的建筑，业主会有怎样的一些要求；领导对哪些方面特别在意，这来自于经验的积累。

刘爱华：您带领的本土设计研究中心[1]是一支非常高效的建筑师团队，作品多、品质高，同时您还指导着中国院其他团队进行着多项设计工作，您用什么样的方法来高品质地把控这些项目？

崔愷：我对设计的快速代入感是来自于多年的积累，这已经变成一种习惯，或者是有一种兴奋感。当看到一个方案，怎么去指出问题，怎么去给出优化建议，对我来讲实际上是一件有趣的事情，我很愿意做这事儿，所以没有太多的压力。

对我自己来讲，指导设计之所以能够快而准确地抓住要点，特别重要的一点是因为我的方法或者说策略是比较清晰的，我知道在这个环境下做这个项目优先考虑的策略是什么，然后再去考虑下一个层面的策略。假如开始做的选择不太对，要通过后面的设计努力去平衡或者弥补，我不太相信这个做法的有效性，我更愿意是大方向的选择是正确的，在下个层级接着做正确的选择。所以指导方案中我会比较明确地判断前面的大选择是否正确，如果是，我就会讨论下面一层的内容。很多时候在对前置条件的判断上，我的作用还是挺大的，因为我的指导而使设计方向得以精准定位，这是让我特别有成就感的。

我给团队提出的意见是带有引导性、创意性的点子，而不是约束性的。规划部门提意见是约束性的，只是指出不符合规定之处，但怎么符合他不管；而我是知道限制条件，去指出机会在哪。虽然我有时会提比较具体的建议，但总体上讲我更愿意去提供一些他们没想到的机会，我会提醒建筑师是不是忽略了某个方面，或者建议应在某方面多想一想，做大方向上的指导和提醒是我作为总建筑师的职责和价值所在。

如果建筑师能够听得懂，他可以去寻找自己的解决办法，我原本是比较愿意这样，但很多时候我能判断出他们是不是能跟我一块儿兴奋起来。即使是本土中心跟随我多年的建筑师，在兴奋度上或者说快速反应上跟我还是有一定节奏上的差异，所以比较急的项目我就会很快画出草图来，不急的项目我就先装在脑子里，让他们先去忙一个星期，之后再看。可能他们找到了更好的解决办法，也可能还不如我一个星期前已经想好的办法。当面对院里有些团队，我就直接告诉结果，比如这么做是不行的，哪些有先天的缺陷，应该往哪些方向去发展。

我常说我是带着价值观来做设计，所以在我提出的具体方法之外可能还有更好的方法，我不认为年轻建筑师要学的是最末端的形式，比如草图上具体是怎么画的，更重要的应该是去理解我说的价值观是什么，如果有更好的办法，我的草图可以改。如果团队不理解这一点，那就只能是被动地接受，把落实我的主意和草图当成重点，而误认为价值观就是个说辞，但实际上我说的比我画的要重要。

从长久来讲，我特别希望通过指导设计把这些道理讲给团队建筑师，如果他们有感悟、也认可，就能在本土设计的价值观体系上形成自己的设计方法、做正确的判断和选择，而不是每次都得把判断这件事交给我。当团队自己的逻辑没建立起来，面对甲方的要求和领导的想法往往就很困惑，缺乏定力，实际上他们纠结的往往是次要层级的问题，主要层级的问题还没解决。如果仅仅按照任务书和某位领导的想法来做设计，那是被动的；主动则是对于这个项目我们的理解是什么，我们代表未来的使用者、未来的市民所做出的判断是什么，这是基于价值观的很重要的答案。在每年的中间思库暑期学坊[2]、中国建筑设计研究院的年底方案评图，我都会讲一些跟价值观有关的判断性的问题。

当选为院士以后，因为拥有这份国家给的荣誉，的确话语权是加强了，但我并不认为我因此就说什么都是对的；之所以现在大家觉得我有很多正确的判断，这是来源于多年的积累和研究。我对设计的对错是非是分明的，所以就很坦然，也不会去刻意地迎合。前些年时常有驻我院监事会的大领导来找我谈话，我都坦率地介绍我的本土设计理念，他们都觉得这些价值观挺好，和政府对建设领域的很多指示都是吻合的。我希望设计团队对"文化自信""绿水青山"这些观念，不是被动去接受，而是自然而然在具体实践中用设计逻辑和执行策略去实现，主动去构想更美好的人居环境。

我们也会碰到确实比较难的一些状况，比如不同领导说法不一致等，团队就陷入到不断的踌躇当中，不知道怎么做。到这个时候，我还是说得再重新思考最初我们是怎么来看待场地、题目以及甲方的要求，要回到"初心"。有些科学家会把保持最原初的好奇心变成研究的动力所在；放到设计上，就是回到最初的判断，当然它不能是过于幼稚或完全不接地气的。怎么形成一个比较成熟的"初心"，确实需要在不断的失败和成功中进行摸索，而经验的积累不太能教出来，这需要每个人自己去磨砺，但是如何能快速地积累正向经验、减少教训，这是可以被指导的。

我认为建筑师应是一个时代的理想主义者，而不能是悲观主义者。现在有些设计师因为自己的理想总是实现不了，所以变成悲观主义者，这时可以反过来分析一下实现不了的理想是不是适合这个时代？要走在时代的前面，甚至带有引领性，这很重要。我对设计还真是一个理想主义者，虽然每次不可能百分百实现，但似乎越来越多的理想实现了，因为我的理想并不是个人美学语言的某种理想，而是怎么更好地处理跟环境的关系、跟社会的关系、跟城市的关系。所以有的项目即使是建造质量差一点，我也觉得理想也往前实现了一大步。

刘爱华：很多与您合作的建筑师都感慨您有超出普通人的充沛精力和快速判断力，这些和您的工作方法有关吗？

崔愷：我不认为我有任何特别天生的优势，我很愿意有机会说一说这事儿。实际上我不是特别聪明的孩子，上学时我虽然学习认真，不调皮捣蛋，但也没有超出老师的期望值或有更主动的思考，我其实比较被动，属于特别怕老师提问题、老师一提问就赶紧低头的那种孩子，做事比较从众，喜欢跟成熟的大孩子玩，不喜欢跟同龄的小朋友闹，比较愿意去学习、模仿。我也不是一个特别善于读书的人，我对记忆力好、学识渊博的人到现在都特别敬佩，我小时候看书、课外读物等倒是看了不少，但是让我具体准确复述书里的内容做不到。到后来学画，临摹一些优秀的美术作品，我也不太能临得特别像，总是差那么一点，但又常找不出问题所在，这实际上还是缺乏特别敏感、发现问题的能力。现在我平常也翻看很多书，但基本上属于留下印象的翻法。所以我写文章不太会用引经据典式的写法，只有当阅读对我产生了影响，在脑子里形成了自己的观点时，我写出来东西才顺，才不会犹豫。这些是我个人成长当中真实的优缺点，我完全不是那种拥有过目不忘、最强大脑的人。

但自从学了建筑设计以后，我觉得自己逐渐走向了一条良性思考的道路，受益匪浅。刚参加工作时，虽然我也算是较优秀的学生，但同龄人之间差距不大，都没有经验，可是在随后的多年工作中，有的人可能觉得遇到的磕碰比较多就逐渐消极了，常用的设计手法还是上学时候学的那几招，并没有不断观察、与时俱进。但我属于不断地在工作当中学习、吸取经验的状态，早期时可能做一个方案磕磕碰碰也过不去，但后来逐渐越走越顺，思考得越来越清楚，每次做出判断也都很有成就感。实际上成就感会让人变得很愉悦，当遇到设计任务时，会有一种非常兴奋的感觉，就不会怕。如果老怕一件事，面对项目中各个条件、各方的不同意见，脑子就会乱了，想不出设计对策；而对我来讲，看了相关的几个线索，在脑子里就很快把它们编织起来，知道应该大概怎么样去解这个题，我的兴奋和愉快是来自于这里。实际上我做设计之外的其他事没有那么愉快，有很多不足的地方。

要说在设计这专业上有什么过人之处，我觉得自己在创作中常会有一种爆发力的形成，这种状态是挺有意思的。记得有一次我和院里两位年轻建筑师一起去看场地，回程一路上我们一直在讨论怎么保护山、怎么去调整城市轴线、观江景方式等，到了酒店我让他们画画草图，他们俩谁都不肯画，可见找到了问题却还没找到解题的方法，所以画不出来，或者怕画不对挨批评。于是我就拿起笔把所想的画了出来，因为这一路上的讨论已经促使我形成一个大方向上的策略，这个策略不是从概念到概念，而是产生了设计的逻辑，草图画完后整个设计的路径基本就形成了。

我今天刚好在"观点"公众号看到有建筑师说"没有争议的观点是没有价值的"，我对这个说法不是特别认可。如果解题的方法会引发很多争议，可能也有创造性，但我一般不往这个方向去。我通常愿意去寻找大家容易形成共识、没有太多争议的解决路径，这是我在价值判断当中特别重要的一点。虽然有些时候我也会觉得我的作品在建成后并没有在网络上得到特别多的赞叹，当然也没有太多的负面评价，是属于比较平和的；但当建成了很久，忽然有人告诉我，说偶然经过哪一座我设计的建筑，还真挺好，我就特别高兴。我觉得这是我需要的一个状态，我并不希望设计的是一个语不惊人死不休的网红建筑，我不是这样的追求，所以我的判断也从来都是在有大多数人共识的基础之上。但确实，如果要产生特别伟大的或者是最有创造性的艺术作品，可能我这个切入点不是最好，没有引起争议，当然也没有引起很大的关注。但是我觉得对建筑设计而言，做一件正确的事可能更重要。

刘爱华： 对于大部分建筑师，当面对某种类型的设计任务，相关工程经验积累是获得专业性的重要条件；您作品众多、方向多元，在不同建筑类型的切换中，如何做到摆脱对类型经验积累的依赖，抓住核心问题，做出高品质设计呢？

崔愷： 我做的设计确实涉及过不少的建筑类型，包括早些年没做过的机场和医院设计，现在也都做了。在本土设计方法的总结中，我特别强调一点——"生活引领"，北京建筑大学的老师在帮我整理相关课程讲义时，还曾特别质疑过这个条目，觉得这怎么会算一个方向。而我认为在设计当中，生活的变化确实引起建筑功能的提升，带动空间的拓展，它其实是一个特别重要的引子：生活是引领，其含义就是建筑是服务于生活、创造更好的生活。我们今天看到，很多原来固有的功能都在向更开放、更融合、更混合、更愉悦的方向去发展，当然功能的基本逻辑、行为模式不会变，但不再是局限于其本身，而是主体功能之上能加入什么，现在变成了功能拓展的一种方法。

比方说图书馆，原来都是设置面对面阅读的桌子，后来发现更愉悦的阅读方式应该是一个人，不是对着别人，而是对着风景，甚至也不见得非得坐在桌子边，还可能是坐在台阶上，或者是躺在沙发上。阅览的方式越来越舒适化，越来越个性化，因为不能说只有坐在桌前的阅读才是收获最多的，也可能是当躺着看本书的时候，人们在最舒服的情况下常是最有感悟的。

当看到人们对生活舒适化、个性化提出来需求时，建筑设计就要跟进。功能对我来讲虽然是熟知的，但是我们不必被过去对功能的定义所束缚，而更要关注到人们行为方式和功能拓展的需求。

我对这些功能拓展引发的创新是很敏感的，尤其是在参观考察当中。比方说医院设计，我前些年到新加坡先后参观过两个医院，人性化设计做得非常好，不仅处处为病人着想，也要考虑大夫、病人家属；不仅是为了看病，还包括怎么到达医院、怎么等候、怎么离开，所有这些都结合医院的使用人群特点有针对性地考虑。

新加坡医院规模并不大，但细节很感人，而且所有设计和管理都是特别人性化、具有开放性。比如医院门口停车接送病人的设计是要让病人坐着等车，门旁还放着轮椅和雨伞方便借用；药房也是多种方式，一般的药是采用小超市的方式售卖，处方药才是药店里卖；医院里有麦当劳等对外餐厅，菜单上标识出摄入的热量，餐厅旁有庭院，也有儿童中心，小孩买完吃的就可以在旁边写作业。我们通常会觉得医院是被病菌污染的环境，但当把绿化、庭院、卫生都做得很好时，医院就变成了社区，小孩可以在这里做完作业去看看住院的奶奶，然后再回家，这在我们的医院里都是想象不到的。让我印象很深的还包括透明的化验室，患者交进去要化验的样本后，能看到在哪个台上给做化验，医院要求医疗内区要向病患公开，以建立信任感，就像咱们看到开放厨房就可以知道卫生条件怎么样，里面的操作人是否认真、有水平。甚至有一些门诊手术室是可以开放，有窗子可以看到里面。这是新加坡医院的理念，特别主张透明性、开放度。

所以建筑师应该把自己代入到使用的情境下，而不是仅满足于把医疗工艺洁污分离和工艺流程了解清楚。实际上，工艺这部分我特别依赖专业团队的经验;而怎样能把人性化和地域性贯彻到设计当中,这是我对项目能做的贡献。

昆山西部医疗中心是我第一次涉及的医院设计,初期介入是原方案中标后业主对立面不太满意,让我做立面指导,我看了也觉得和全国医院建筑都差不多,这类建筑都不怎么好看。我建议不要单纯改立面,而是去考虑怎样能够把江南园林文化渗透进去,让中医有传统文化的环境氛围。江南园林的做法本质上是在房前屋后的紧凑布局中去实现小空间的渗透,而并不依赖于多大的一块绿地。在一个很长的完整立面上去表达江南水乡的意境是很难的,没有适宜的尺度感,所以我建议把门诊楼切成一段一段的,这样院子和墙都有了,顶也就可以有了,同时还设计一条长廊,让病人可以休息,也可以排队等候就诊。

另外,几乎所有医院前面都是停车场,我特别反对这个做法,医院前面一定要是一个花园。正好昆山医院要做一个大型地下车库,还有建设中的地铁,所以我们考虑了很多地下空间和地铁、停车场的流线衔接,做到地上无车化,形成了园林的前庭。

我们还将候诊区做成内外走廊结合,放上座椅,大家可以坐在这儿候诊取药,或者等手术、化验结果。昆山医院是一个很大的带有休养性质的中医院,很多病人是来自周边上海、江苏的慢性病患者,他们会比较长时间住在这里,也因此会有很多家属到这里来,所以要考虑给患者和家属提供适合的场所。我觉得所有的设计都应该是考虑患者在生病的状态下对空间的特别需求,让他们能得到一种关照和享受,所以我们在住院楼上设计了每床一窗、每层一厅、带遮阳棚的阳台,还有屋顶休闲的花园。

这一系列对使用的考虑同时跟江南的园林文化结合起来,使医院有了地域性,这是以往医院设计当中比较少考虑的。完工使用之后,中医院的院长特别满意,而这次的医院设计对我也是难得的一次经历。

机场设计也是这样,我最初接触到机场设计是从厦门机场开始,这是院里的杨金鹏建筑师的中标项目,请我帮着去做设计上的一些提升,从此之后当我出差时就会更关注机场。我们做的济南机场是委托设计,方案有许多创新点,我记得当时山东的领导说,虽然崔院士设计机场不多,但是坐飞机很多,他是以乘客的身份来做设计的。对我来讲确实就是这样,实际上我每一次经过机场,都会仔细地观察。有的机场花很大的代价做屋顶,造型是很好看,但如果是下大雨时去机场,因为落客区只有很局部的地方是能遮雨的,在其他地方下车全都会被雨淋,所以大家只能在车里慢慢排队等,这种设计是很有问题的。

所以在济南机场设计中,我认为立面造型意义不太大,而特别重要的就是怎样让落客空间品质高、功能性强、让人觉得愉悦。现在很多机场的屋盖挑到很远也没能解决遮雨的问题,后来一个流行的方法是在底下再加个小棚子,厦门机场就是这么做的。等于是建筑没解决问题,最后由运营方来解决,这不是好设计。因此设计济南机场时我说一定要先从这大棚子做起,遮雨的同时又让大家感受到阳光和风景,晚上加上灯光也很绚丽。然后再分析大厅的空间重点在哪里,哪个地方要低,哪个地方要高,而不是像现在很多大型机场都是泛泛地高大,其实没有必要。有的地方稍微矮一点会让人觉得有一种亲切的尺度,有的地方高起来又感觉到空间的立体张力很大,空间变化的同时,也兼顾了节能方面的考虑。

我一直认为设计是有对错之分的,而不仅仅是因为大家都这么做所以很多问题就被忽略。比方说航站楼里边怎么样看飞机,很多人都觉得这一点无所谓,反正登机口那儿能看飞机,而我认为旅客进到候机大厅后就应该很快能看到飞机。我喜欢首都机场T3航站楼,每次早上赶飞机,从车道上过去的时候,能看到最端头很窄的空间,透过很透明的幕墙可以看到飞机在后面,商务办票口在这里,办理登机时会感觉到自己是在机场的空间中,而不是在一个封闭的大空间里排长队。要让旅客能够迅速地看到飞机,这就造成内部空间布局上的选择,包括哪些地方不能放办公空间,应该要把它跟旅客的流线和视线方向顺起来,不能挡。现在很多小机场,并没有那么多客流量,但柜台一拉一大横排,反正旅客们最后总能找到某一个柜台有工作人员,到那儿去办票,然后越过安检才看到飞机,这个做法实际上很不好。

还有机场餐饮区的设置,最常见的做法是放在人行通道的两边,店面就像一个个街边店,方便是方便,但是餐饮本身就没有景观的价值。实际上在机场吃饭,虽然有些时候是很匆忙,吃完一口就走了,但有些时候要等飞机,可能要在这儿坐更久一点,所以设计中我们会去想怎样能够让餐饮跟空侧建立视线上的联系,尤其轻餐饮,应该有很好的位置。

林林总总的许多从旅客感受上的考虑带动了我们对功能本身的完善,从大的功能布局到细节的功能都用人性化的方法去提升。这应该是建筑师的一个职业习惯,只要想到人对使用建筑的体验,任何类型的建筑都会有近乎相同的判断,但结合不同的功能又有不一样。支撑我们跨越不同类型的项目并找到设计的亮点最行之有效的办法,其实就是找到那些功能流程背后的那些最基本的人的生活行为特征,并对某些司空见惯的做法进行反思，找到需要改进的方向，由此破题。

比方说像医院的诊疗室没有外窗现在是很普遍的做法，可我记得小时候去家附近的公安医院，当时也不算小医院了，诊疗室都有窗户的，大夫看完病以后还能在窗边喝杯茶。但现在的大医院设计都是大平面，根本开不出这些窗，造成医生长时间、高强度地在封闭空间中工作，这很不人性。但因为做到了洁污分流，也都有空调，很多人觉得解决不了开窗也无所谓。而在昆山医院的深化设计中，我要求团队一定得把竞赛时没有解决的问题解决掉，不能光是照顾病人的感受，也要照顾医生。一位分管副市长在有一次听我汇报时，也曾专门问到：病人来医院是短时间的看病，而医生在医院是长时间的，那么医生的工作条件怎么样？在哪吃饭、在哪换衣服？我就专门给他讲了一下我们的设计。但当时医生食堂还是一个封闭空间，后来我们专门调了这个空间，我说市长说得对，医生那么早来上班，吃早饭时一定要看得见阳光和景观，才能心情舒畅，而不是在封闭的空间里匆匆吃了饭就走。

这些调整实际上就是一种比较全面的人性化考虑，人性化绝对不是口号。我觉得对设计师来讲，如果提出一个口号，那一定得是认真的、要去执行的，千万不要养成说空话的习惯，如果习惯说空话，就不是一个认真的设计师。你不能跟人说一通大道理，但最后做的并不是这样，那就是没有把理念和设计结合在一起。

刘爱华： 在设计中，常会面临业主的固有理念和创新之间的矛盾，创新既有来自建筑师的创新价值观，也有来自业主对创新的期望，您如何看待并解决这个矛盾？

崔愷： 如果业主自己有创新上的需求，建筑师应该去充分地了解，而不是自己另找不同的创新方向。我每次听业主的诉求时都是很认真的，我能听得进去，不是被动地接受。哪怕业主观点里有很多讲得不太对，或者更是从经济上、成本上、自身利益方面上出发，但是我会关注其中的积极因素，然后看能不能在这个方向上比他想得更好。顺着业主的思考往前想，而不是往另外一个方向想，这是我处理这类问题的一种习惯。这样的话，我们要解决的问题跟业主想解决的是同一个方向；当我们拿出更好的解决方法时，业主会觉得物有所值或者是超值。很多次业主在听我们团队汇报完方案后会感叹说大师就是大师，言外之意是我们给出的方案比他们想要的还要好得多。不忽视业主意见，也不老想着去说服业主，而是怎么样帮助业主想得更好。

帮助业主想得更好，不仅仅是指超值满足业主方具体的利益需求，还包括怎样引导去把公共利益融到设计中。实际上这需要有很细心的拿捏：这个项目中可能带来的公共利益是什么？怎样在不影响业主基本利益的同时能满足共同利益？因为一个项目在其业主代表投资方利益的同时，还有一个重要的审查方是政府，而政府会更关心城市利益、公共利益，所以我们在设计中重视公共利益的思路和判断，实际上这是业主不太关注的，他们会比较谨慎地反复打磨任务书，其他只要满足规划条件就可以。但当我们拿出的方案不仅解决了他们的诉求，还兼顾了城市整体利益的时候，业主方会感到释然且有成就感：项目在政府那里容易通过，投资方利益保证的同时还找到了与公共利益的平衡点。因此我们比较少在方案优化过程中反复修改，如果仅仅去满足业主利益，然后政府、规划部门不同意，然后再按政府要求改，业主又不同意，这样来回"拉抽屉"的状况是有些建筑师经常遇到的，甚至到最后把原方案的构想全改没了，完全失控。我们是主动地把这两个利益想在一起，这是很重要的，也因此经常得到业主和规划管理部门的赞许。

刘爱华： 在实践项目中，您是如何处理结构、设备技术、景观室内等其他专业的问题？

崔愷： 一个成熟的建筑师要能够驾驭建筑当中所有的专业，与各专业工程师建立积极互动的合作关系能很好地引导和管控各专业的技术方案，甚至能推动各专业把建筑设计做得更有技术上的创新性，这是我这些年的认知。

原来我对这方面的认知还仅仅是怎么样让这些专业服务好建筑设计，当然这首先得了解这些技术设计的基本原理和规律、它占有的空间形态、大概的设备系统特点。这些我都是比较早就很有收获，在我刚参加工作没多久，就做过管道综合、画过一些技术性图纸，当这些画完以后，这些内容真的很长久地停留在脑子里，比方说走廊断面，我会很清楚地知道这些管道应该怎么排得更合理。掌握并把它变成你跟设备、结构工程师讨论问题的知识基础，这是很重要的。否则只好由其他专业来分，就会比较被动，有些年轻的建筑师在这些方面经验不足。所以我在指导设计的时候，实际上不完全是指导概念，很多是在纠正设计错误。

在最近的西安碑林博物馆深化设计中，团队很认真，也深化了很长时间，有天拿出几个问题说再跟我碰一下，我一看还是挺大的问题，原设计中在沿着主要的街道这侧都是有商店的，现在全都变成风道了，墙上到处都是百叶窗。我说应该早就跟我商量这些事儿，风道放在这里合适不合适？虽然现在甲方似乎不太在意，但是我们要考虑提供未来能够经营的场所，这是一个城市空间的经营。不能说这都是风道，商店就没地儿去了，一定要想办法，比如风道能不能垂直于街道，我提出来这些建议之后设备他们还真就解决了。还包括有的项目审图时我发现车道中间有柱子，我上下对了几层，看出这柱子是可以拔掉的，结果果然可以拔，那之前为什么不早点想着拔掉？包括卫生间的设计，都要认真考虑怎样能够能排得更有效率，入口空间视线"不穿帮"等。这些小地方也是我常常比较敏感的。

成熟的建筑师眼睛一定要有这种洁癖，这些细节其实都是建筑设计的质量，要能够为客户主动着想，把设计当中的错误减少到最小，这是一个成熟、优秀的建筑师要具备的基本素养。更何况，我们现在要做绿色建筑，要主动地引导其他各专业去适应这样的大方向，关注低碳节能对结构、对设备技术的要求，我觉得这也是一系列的创新。我不认为创新只是一个所谓的概念创新或是形式创新，它应该是一系列内在技术品质的创新。

现在行业里有一种比较不好的制度，就是做方案以后重新招标做后续工程设计，初步设计和施工图常是由总包的EPC团队来做。我每次说到这件事时，都跟业主掰开了揉碎了讲，设计的创新和高质量完成是依赖于从设计方案到施工图甚至到施工配合当中一系列的主动积极的应对和调整，要切开了，后边那段一定没办法高品质完成任务，他们只负责外形方案不走形，里边可能就随意调整拼凑，怎么省事省钱怎么来，因为他拿到这个工作并不具有创造性，所以会安排一般的建筑师帮着完成，遵守规范没毛病，保证审查通过，可里边的品质是没法保证的，什么都是靠装修，更不要想什么技术创新，很多创新机会实际上就错过了。

所以我每次都强调从头到尾的全过程创新、全过程高质量设计，这是特别重要，这就意味着要对其他专业要有一个持续的统领和指导，也包括室内设计、景观设计、照明设计、幕墙设计这些专项设计，现在几乎每个项目都得过一遍或几遍。常年跟我们合作的王东宁总，他是非常优秀的照明设计师，获过很多国际照明大奖，但每一次他听我讲都觉得很有收获。景观设计等也是一样，听我讲一讲这个项目的来龙去脉和我的一些思考、基本价值观，可能会减少设计上走弯路，而不是一上来就按照套路做，那样的话肯定与设计整体思路脱节。

对合作的各专业，建筑师不仅要有一种主动设计的参与感，在参与中还要主动把建筑的理念贯彻下去。虽然我也知道有的建筑师不太能贯彻下去，或者业主、造价、施工单位等各方面条件不允许，但是总归别出大错。比如说我们仅仅是建筑带领着结构设备专业能做到80分，如果接着控制室内、景观设计就可能做到90分，如果再把幕墙做好了，可能到了95分，但门把手和楼梯扶手做得不好，可能剩下这5分就丢了。但是还是要每一次能够到什么地方尽量往前够，有些简单的项目基本上能够到99分，但是有一些项目就会缩回来一点。但如果整个体系性控制好，大概就能到90分以上。所以我觉得建筑师作为设计主持人，应该有理想主义，努力争取怎么样能接近100分。当然这还只是设计和建造，实际上最后的使用是不是能够对也很重要，像南京园博园虽然碰到疫情，但是使用管理得特别好，今天上午董元铮建筑师还跟我商量，业主说酒店运营反映原来多功能厅太小了，满足不了大型婚庆的要求，提出要不要再做一点扩建，我们当然高兴继续配合。像这种使用状态特别好的就能到了110分、120分。

随着成熟度的提高，我视野看到的东西、观察的东西、记住的东西很多，所以在知识层面上就可以跨越专业的划分，将设计变成一个泛设计。包括平面设计、标识设计，我们也做一点。前段时间泉州火车站要做一个雕塑，问我能不能做，我开始说不行，团队的任祖华建筑师准备找美院的朋友做。我想一般艺术家如果对环境不了解也会做得不得要领，而福建民居大屋顶的脊和瓦都很漂亮，可以看看怎样用土建材料或金属结构把传统的"瓦"和"脊"描述出来，我按这个思路画了草图，后来就这样做了，甲方很喜欢。不敢说从纯艺术角度来讲有多高水平，但这是来自我对地方文化的学习和体会，因为实际上车站广场上应该表达的是地方特色，而不是一个纯艺术的造型。

刘爱华：您参与过多个重大项目的设计竞赛、投标的评审，作为评委您是如何评审参赛作品的？

崔愷：评审确实是一个既是考参赛团队，又是考评委的一项活动。我不认为评委拥有非常任性、自主判断的权力，可以随便来评价别人的方案，所以首先要尊重设计团队的努力，同时也尊重业主对建设这个项目的期望。通过我们的评审去得到一个相对比较好的解决方案，这是我每次评审怀有的基本想法。

我在评审中能够迅速地找到问题，或者说思考投标项目要解决的问题，这得益于平常在一线快速、大量的设计工作和经验。从这一点上讲，就有点像把自己代入其中，如果我是设计方要设计这个项目应该怎么做？而不是像一个局外人，拿以往一般性的规划指标或一些设计要求的框框去套。

我会很重视设计方案中的环境分析和最后的设计结果，中间属于理念陈述的部分听一听就完了，如果时间短我基本就不怎么看，因为很多理念大家都司空见惯。重要的是怎样把理念贯彻到设计当中，这实际上是看对场地分析和设计结果之间的合理的逻辑性。评委不会被泛泛的、观念性的东西所感动，而是更去看设计结果。我对高质量完成的设计成果有先天的偏好，哪怕方案遗存的问题比较多，甚至有争议，我都觉得还是值得重视的；有些竞赛设计的表达完成度比较差，让人觉得即便它有哪个点踩准了，也很难真正做出一个好的设计。

但同时，在很多评审尤其是城市设计评审中，确实牵扯到价值观。比方说这个设计本身要解决的问题是一个中小型城市的问题，设计团队却用一些世界著名大都市的参考案例来作为成功经验来说明设计理念，这完全是脱离实际。所以我更重视这个城市它要急切解决的问题，它在发展上的可能性以及它应该呈现出的一种定力或自信，而对"忽悠型"的设计很反感。竞赛虽然是一个设计的开始，但是设计团队的立场和介入的方法实际上是很重要的一个判断点。

再后面的一个层级是看设计成果综合表达的品质和意向，是不是恰当地反映设计任务的需求，是不是太夸张或太过度用力，不能仅仅是外观设计，要同时能看到很多内部空间的设计。水平比较差的团队通常用力都在外表上，一看平面和空间，什么都没有，这类是我最先否定的方案。如果设计的外表有点夸张，但这个夸张带来了一个非常好的空间体系，我觉得也是很值得欣赏。徒有其表的设计我是坚决反对的，不仅不能徒有其表，而且空间还要能反映出这个地区的文化特点，功能上有创新性的引导，还有公共利益的考量，比方说虽然做的是一个建筑，但是考虑了开放性、跟周围城市环境的关系，这些都是很积极、有价值的地方。

总体上我在评审中是用几个层级来看：第一，问题分析得准不准；第二，设计的深入程度够不够；第三，是不是徒有其表的建筑；第四，设计是不是能够呈现出设计团队所追求的价值观。

以上这些关注点，我在评审看图时就很快能做出判断。我们也要听参赛团队讲，我在听的时候会记一些笔记，但还是要看图，哪些方案不行就不再细看了，哪些方案还行我再仔细看；哪些方案好，我就拼命看、看仔细，再挑一些小毛病。所以我经常在评审意见上写的是带有设计优化建议这样的意见，而不是简单说哪个好、哪个不好。对于入围或者前几名的设计方案，我会特别乐意充分肯定它的价值，这些肯定的评语甚至会成为甲方向更高层领导汇报所需要的支撑。因此还有过几次让我代表评审评委跟业主或领导做汇报，因为觉得不仅我的评语写的好，当场说的也好。

我认为评委有代入感是很重要的，这并不是说要求所有评委必须是由一线工作的建筑师来担任，但如果评委自己也做设计，他就会设身处地地去想问题的解决方法。我在深圳参加过很多学校设计竞赛的评审，当时有几个年轻的建筑师评委，像董功、孟岩，我觉得他们看得更细、讲得更细，也一针见血。所以我觉得优秀的设计师当评委是对项目很重要的一个保障，能把好的设计推出来，也能把次的设计屏蔽掉。就怕评委的眼光差，如果再带一点私人或相关方的利益，就经常会做出比较差的判断。所以竞赛评审是非常重要的，我确实也因为有些时候对评委水平的担心，有些设计竞赛我也不太愿意参加，因为觉得设计的用心、对场地的判断等，有的评委是不在乎的。

刘爱华： 在这个信息爆炸的时代，面对海量的书籍杂志和网络资讯，您是如何选择阅读，并与设计实践相连接的？

崔愷： 实际上我学建筑，阅读不是我最重要的获取信息的方法，确实要用到书的时候我会去翻，但不是系统性的、有计划的那种读书，看网络信息也比较少。但是我特别重视亲身的经历，这么多年来有很多机会到国内外考察，这对我来讲是受益匪浅。我今年也是在我夫人的帮助下，把历年来考察建筑的照片做了系统性的整理，使用起来很方便。包括最近给北建大做本土设计课程讲义时，要挑国外的作品案例，选的都是我去过的，这样我就能讲得详细深入。再比方刚才我们聊到的医疗建筑，我曾2012年参观新加坡邱德拔医院，它的很多细节考虑对我们这次的昆山医院设计都有影响。

所以面对大量的资料，重要的是怎么把它们在设计中用起来，我想大部分建筑师也都有这些资料。我每次看这些东西时候，实际上我是进入到照片当时的场景，我会有不一样的阅读。所以我特别重视个人的体验，这是我提升自己设计的一个很重要的方法。

像《建筑学报》《时代建筑》《世界建筑》这些主要的建筑杂志，我也会翻看，就好像去看参照系，看到同行优秀的作品对我来说是一个鞭策。当然在看到有的作品整体很好的同时，有些地方是我个人觉得过于较劲、过度设计，这种情况也是有的，我在看的时候会有我的判断，哪些地方可以采用更适度的方法和尺度来去解决实际的问题。

所以个人的体验和记忆、带有观点的阅读，这两点对我是很重要的学习方法。

我最近在听余秋雨在文化书院讲的课，听了有两个月了，每天走路时都听。我觉得特别好的一点，是里面对中国文化的好多场景描述让我特别有同感，它不是那种从不知道到知道的知识性学习，而是一种引发共鸣式的学习。它让我觉得我的思考、我的工作，包括我对价值观的判断，都在中国的大的文化的语境之下，这让自己越来越觉得活得挺明白、干得也挺明白。

刘爱华： 每次本土的季展上，您都会为展出的作品题一首诗；还有上次看到您的草图本上，在南宋皇陵遗址博物馆的构思草图旁边也写有一首诗《向北方》。为什么会用写诗来表达设计上的一些想法？

崔愷： 我在小时候不知不觉养成了一些文学上的喜好，也写一点打油诗。实际上写诗这件事对我来说是用比较简洁的语言归纳比较复杂的事情的一种习惯，当然要建立在有情怀触动的基础之上，没有情怀肯定写不出诗来。而语言的凝练实际上来对设计上的判断。比方说为什么我会写《向北方》，在做那个设计时我会去翻看一些与南宋时期相关的书，正好我一位同学曾送过一本研究宋代城市格局的书给我，但我一直没怎么读，然后那天我翻了翻，看到说在南宋临安大部分的建筑，包括城门，都是正门在北边。这个方向感引起了我的注意，因为我们太习惯于坐北朝南了，什么东西都在南面，到了南宋为什么会向北，忽然理解到这其实是来自一种他们要回到故土的乡愁情怀。我觉得这就是设计需要表达的情感，所以一下就形成了用"向北方"这样诗的语言体系来组织，抒发情感的同时也是在组织我的设计，这首诗是设计思考的一部分。

这个项目在绍兴附近，我去看场地时，一路晒着大太阳在茶田里走，陪同的人介绍说现在场地上什么都没有，但在当时每一位南宋皇帝的墓上面都建有一个漂亮的房子，棺椁都是从临安用船送到山口再抬进来。这很有意思，跟我大学时测绘过的清东陵，包括与明十三陵的感觉都是不一样的。通常情况下皇帝一上任，就开始寻找自己的墓地，然后修建它，是把修建陵墓当成自己人生的一部分，它是君王的领地，要有入土为安的感觉。但是在绍兴南宋皇陵这里出现了死后不能入土的情况。当我把这些情怀以及顺应的设计思路讲给甲方时，他们觉得很有意思，原本他们觉得场地上什么都没有，真是不知道怎么做。

原来我对历史并不太感兴趣，后来听余秋雨讲不同时期的文化还是有很多收获，魏晋南北朝是一个文化的高潮，唐朝诗人虽然很多，但是实际上排在前面的不过是李白杜甫这么几位，而陶渊明是魏晋时期的代表人物。正好我们在做重庆酉阳的一组文化建筑，酉阳据说就是桃花源的实际场景所在，那里有巨大的溶洞、美丽的田园，也有很多对应的历史记载。所以我在设计时，联想到陶渊明落寞孤独同时又志趣高雅的文人情怀，会有一种在用设计跟他隔空对话的感觉。

以前，建筑设计对文化传承的表达经常挺被动的，只想着琉璃瓦、大屋顶那种过于强调形式感强但又特别受到束缚的模式，让人联想起民国时人希望回到清代，希望重新把皇帝立起来。我们年轻时对这些仿古建筑都非常不理解，也特别不感兴趣。像戴念慈先生做的中国美术馆、张镈先生做的民族文化宫，我觉得在这个方向上他们是做到顶峰了，叹为观止，后面没有人有能力做得更好。

所以我们应该转向其他的方向，怎样离开形式感去做文化传承。像老子讲的话至今仍然是中国建筑学的经典，国际上很多学者也都知道，我觉得用这种隔空的文化感悟来表达文化的传承是一个更广阔的空间。现在有不少优秀的建筑师在设计中研究和借鉴山水画、文人的情怀以及有特色的生活文脉，去找文化上的共鸣，这些对传承的思考跳出了上几代人没走出的民族形式现代化的怪圈。

刘爱华： 您喜欢读的书里面，更多建筑专业领域内的书还是更多专业外的书？

崔愷： 建筑专业内的书，有一些书我会读一遍或两遍，但是并不太多。我最近读的比较多的是一些小的传记或者是像建筑师们写的回忆录、纪念文集，我倒还蛮有兴趣的，主要是看那个时候的历史，看人在那个年代那种理想主义，看他们的思考和困惑，从而更多感悟人和社会的关系。我觉得到了我们这个年纪，就会想把自己这些年来的忙碌、积淀下来的思想跟前人做一些并列的比较，去看当年梁思成先生怎么想的、卢绳先生怎么想的，他们怎么处理当时遇到的事情，他们留给我们后人的启发是怎样。专业外的人物传记我也挺喜欢，以前也读丘吉尔、奥巴马、乔布斯这些。最近也在看《简·雅各布斯传》，一个社会工作者对城市的观察、呐喊和斗争。

这类阅读有点像从做人的角度去看待我们今天做的事在历史上的某种定位，这说起来有点大，但我觉得建筑师作为一个时代的很重要的建设者，确实第一应该有理想主义，第二应该有对历史负责的态度，能够有更宽的视野，做的设计让城市更好一点，让乡村更好一点。虽然我们也知道建筑师能做的也很少，但总还是要怀着这种大情怀去做，哪怕做一些小事情，我觉得这还是挺重要的。不能说因为做不了大事情，对现在的社会现实中的问题很不满意、有很多看法，就不去响应社会的需求，躲起来做自己的，这叫避世，我不是特别喜欢。佛教、道教、儒家之中，我个人觉得儒还是很好，佛教也还可以，道教个性太强了，完全不能容忍去融入社会，满足于自己的逃避，这对社会没什么好处，但是道家的思想对社会还是有好处的。但是想象老子这一辈子就留下这五千字，这五千字是很好，但他原本可以留下几万字是吧？李白就留下了那么多首好诗。

所以对我个人来说，我并不特别关注自己的哪一个作品是最好的或者哪件事是最有意义的、觉得其他的作品就无所谓。有人是这样想，这个机会要把握好，做一个大作品；那个工程希望不大，就做个赚钱的活儿吧。我想这种机会主义的立场对城市是有害的，对建筑师自己也是不太好的，扮演了一种"变脸"的角色。好的作品拿出来反复讲，不好的就不提，甚至不好的或一般的作品比例还远远大于好作品，你说这是个负责任的建筑师吗？充其量只能说是个有才华的、聪明的建筑师。我对自己的要求是对每一个建筑工程都要设计做好，要对得起这个机会、这片土地，这个事特别重要。有些时候别人也会问我为什么做这么多项目，多也是我们的贡献，好的建筑作品多就意味着为社会做出了更多，忙并快乐着，也因为有这种理想主义的心态，反而对自己的要求就更高一点，尽量把设计做得好一些，在这个过程中保持进步和探索的状态，而不是故步自封、自得其乐。

刘爱华：您是如何在职业生涯中长期保持一种对社会的敏感度和对创新的新鲜感，突破固有的认知，不断获得自我更新的能量？

崔愷：对建筑师这个职业，我觉得持续的学习是一个基本要求。因为这个职业是服务于几乎所有的行业，生活中绝大部分的空间都需要建筑师去设计，而建筑师在学校里学的东西实际上又是有限的，学的是基本功，甚至可能在学完以后，有些工艺流程、建筑特色等这些可能在时代发展中已经出现了根本性的变化。国外对建筑师要有继续教育也是因为建筑师的职业要求，实际上并不是所有的职业都有继续教育的这种行业性要求。不断学习是建筑师的职业性要求，这是第一点。

再一点，中国院给了我这样的一种站位。我刚到院里工作时，看到那么多有名的建筑师在这儿工作，对自己的要求就跟去到一个小的设计院是不一样的。之后又看到这么多优秀的同事，那时新入院的大学生竞争力也都很强，所以对我也是有一种自我鞭策，希望自己能够在这好好做，所以当时对一些社会上的干扰比如下海等，都不为所动。

同时，在参加建筑学会活动等行业交流中，当时很多老先生对我们年轻一代的激励、鼓励和关照，也潜移默化地成为自己的一种动力。虽然我有些时候说不太准，到底哪些老先生在哪些事情上真正教过我，当然从我的导师彭一刚先生那里肯定是有很多很具体的收获，但在这些学术交流当中，耳濡目染看到的更多的是文化素养和精神。

在时代的进步中，建筑设计也不断地创新，尤其是2000年以后，国外建筑师进来开阔了我们的视野，包括我们不断出国所看到的差距，包括海外学子回来以后的建筑创作起点都很高，他们有着更多的见识和国外教育体系下所收获的理论性和实践性的成果，这些都让我觉得这是一个高度竞争性的行业。我并没有因为我是77届，就能够摆老资格，就觉得这代人中我们是算老大，我并没有这样的感觉，而是时时觉得行业进步在推动自己要不断学习，这些都是我背后的动力。

当然，在找准本土设计的方向之后，我也觉得自己在创作思维上进入到更加理性、更加有自信、更加清晰的一种路径当中，作品也系统地提升了水平，成功率也比较高，在这种情况下要保持更高的追求，而不是不断地重复自己。所以在本土中心我们提出了一些技术发展上的研究方向。比方说清水混凝土，在实践项目中应用高难度的清水混凝土，并从设计上去解决相关的技术问题；比如木结构，对木结构的尝试原本是来自一个偶然的提议，但竟然做成了，那么现在我们就会创造条件继续实践，因为木结构在绿色低碳方面实际上是一个特别好的材料，是碳汇材料；再比如轻型结构，原来我们一直抱怨"肥梁胖柱"，这源于是结构设计上的保守，建筑师也不太参与，我们现在结合新的项目创造轻钢结构，或者是轻型结构、轻量化结构；还包括现在做装配式，我们也有不同的思考和实践探索，得到了业界的高度关注。我们在不断地找一些阶段性的目标，去关注技术，关注绿色建筑创新，这些都是特别主动的。

有些时候我翻《建筑学报》，以前觉得我们自己包括中国院的作品都做得挺不错的，其他有的作品就做得不太行；现在翻开学报介绍的作品都做得特别精彩，很值得学习，有时间也想去参观。所以如果我们仅仅在一个小的范围内互相比较谁做的好一点或者谁的辈分高一点，有点无聊且视野太窄了。我经常跟院里的年轻建筑师说，一定要有行业视野，我不一定要求大家都有国际视野，但是行业视野一定要有，要知道在同样的中国创作语境下，仍然有一批优秀建筑师在做很出色的实践，从创意到掌握的先进技术、到高完成度，这都是我们大院建筑师是要积极学习的，也是给了我们压力。所以这也是为什么我希望大院要办教学，也督促大家多写文章、参加各种学术活动，我觉得这是特别重要的，要跟得上，不能只看自己身边。这一点特别要嘱咐我们后面这一拨中青年建筑师的，如果想对学术性、事业性有一定的追求，需要自己加码，这个是非常由衷的建议。

我在获梁思成奖之后曾写过一篇文章《在中间》，实际上我描述的是一个不是所有人都能读得懂、但是我内心所想的一种状态。"中间"是很重要的一个位置，就是以什么样的态度去处于学术圈内圈外、大院小院、年轻的老一代之间。在这"中间"，实际上能够学到很多，要保持开放性，既不去刻意地站在某一方，也不会很激进、很"愤青"，更不去自我欣赏、故步自封。很多地方请我去做演讲，说明通过开放性思考，我还保持着创新的状态，总有新的东西可以拿给别人讲。其实在时代转型中，的确又有许多新问题也在要求我们重新思考和面对。总体来说，与时俱进，在时代发展中不断自我鞭策，这特别重要。

注释

1）本土设计研究中心，是由崔愷院士主持的建筑设计团队，成立于2014年2月19日，其前身是成立于2003年的崔愷工作室。下设一个研究室、三个工作室、一个办公室，目前共有正式员工65人。成立二十年来，本土设计研究中心共设计200余个项目，其中建成作品100余项，获得地方、国家及国际建筑奖项100余项，作品范畴涉及建筑、景观与城市设计领域。

2）中间思库暑期学坊，是由中国建筑设计研究院举办的公益培训，从2014年至2023年已举办9届，面向在校大学生、研究生及青年建筑师，由崔愷院士担任学坊主持，来自中国院的优秀建筑师担任导师，聚焦于对北京城市更新领域的关注，通过学用结合的模式，构筑从建筑教育到职业实践之间的桥梁。

DIRECTION AND APPROACH: AN INTERVIEW ON PRACTICE OF LAND-BASED RATIONALISM

Interview with:CUI Kai
Compiled by:LIU Aihua

LIU Aihua:

In this book, you summarized a series of methods and strategies of Land-based Rationalism design in strictly systematic patterns. How do you think these methods and strategies should be flexibly applied in the face of complex and multiple realities in specific projects?

CUI Kai:

When we emphasize indigenous design, we actually emphasize the interactions between architecture and external conditions, which are usually multi-layered and multi-directional, and their impacts on architecture vary. In designing specific project, we can screen these complicated interactions, grasping the main contradiction at the beginning, adhering to solving it with insistence while finding solution for secondary contradiction and making due judgment in the process.

I usually stay open and relaxed in mind at the beginning of a project, instead of deliberately looking for ideas. When I receive a design task, my team will collect relevant information, but I will not put a large number of reference cases in my mind, nor will I be particularly anxious to make subjective judgments. What's more significant for me is to keep fresh and new of the scene. I can obtain enlightening perceptions through walking on the site, talking to the owners, and observing the surroundings. Perception is particularly important for design, in which it will be determined what the major concern is, whether it is a problem brought about by environmental characteristics or a problem caused by the main function.

Now for most buildings, especially cultural buildings, there are basically no problems that cannot be solved in the functional arrangement. But in the context of a specific environment, how to take environmental advantages, to solve or avoid environmental disadvantages, and to form a reasonable arrangement of functions? The relationship between function and environment shall tend to be determined.

For example, when designing a museum, I will judge about the internal and external environmental space, whether it is open or enclosed, quiet or ceremonial. Taking library design as another example, is the external environment of the base scenic or noisy? At which levels or locations the environment is chaotic, and which is quiet? I ask these questions repeatedly in my mind. We know where are the favorite places people like to read and how to find such a place on the site. Therefore sometimes we put the reading area in a high place, sometimes sink it half-underground, using these options to deal with the bad and noisy drawback of the environment. If the site is quiet, what we do is to let the reading environment blend into the surroundings. These judgments will become interesting expressions in the design.

During the site survey at the School of Urban Design at Wuhan University, I actually knew what to do after walking through the site: with a mountain on one side, and student dormitories densely clustered on the other side, the main teaching areas should be placed on the mountain side, while the other side be lined with smaller-sized teacher studios. I even thought about how to hide this part and form a visual isolation from the dormitories. From a vertical perspective, considering the height difference from the mountain down to the site, I thought over how to make use of height difference to the teaching area, rather than just connecting the floors with stairwells, so as to make the space like a mountain. The idea of classrooms placed on a "mountain" was also related to the design teaching itself, forming a tiered teaching space that creates both a sense of domain and sharing. As is the case, I usually make judgments on site, and when thinking through them, I am particularly excited. After the trip to the site, I have come up with a clear plan.

The starting point in design comes from the site, not from the architect's own imagination or inspiration of a certain master. It is different from the approaches of many architects, of whom some have an ideal space in their minds. When considering the ideal space is the priority, and the environment is deemed as secondary concern, the factual practice will be relatively passive. It may not be necessarily unsuccessful, as some buildings turn out to be quite successful, but sometimes there seems to require a better solution for the site.

Of course, there are property owners or managers who do not particularly care about the characteristics of the site we have indicated. They believe that they have right to change the site, and so do they operate. Therefore, if you design with self-consistent logic, sometimes the feedback may not be very passive. As long as the appearance is good-looking, the client maybe goes well with it. But for me, design based on land means a different methodology, so that I often need to convince my clients to realize the importance of the environment and the basic logic of building and site coordination. A series of important ecological protection initiatives, including "green water and green mountains", "handling lightly" in urban renewal, and "leaving behind nostalgia" in rural revitalization, put through by the government, are all directly requiring construction to be related to the external environment. The logic of Land-based Rationalism is to use the initial concept of the building to meet the requirements of harmony with the surrounding environment as the first priority, and I believe that such a harmonious posture will not fundamentally deny the rationality of internal functions. So has this design logic received increasing support.

I referred specifically in the article "About Site Survey" in this book that it is important for me to understand the task before going to site, and then go to the site with questions in mind. Compared to listening to others' introduction, it's better to visit by myself. Because of such reliance on on-site judgment, if I haven't been to site, I ask more questions when listening to the team introducing the environment during discussion, and even ask questions that they are not prepared for or turn a blind eye to. I have encountered this situation multiple times, indicating that the team are not always perceptive enough to the characteristics of the site.

In another case, I went to Switzerland with GUAN Fei to discuss the design scheme of Rizhao Science Museum with Professor Samuel Ting. Before leaving, I received an assignment that the Ministry of Foreign Affairs would build a consulate in Strasbourg, but were not satisfied with the plan made by a French architecture design firm. They asked me to go and have a look. After our work in Geneva, we drove to Strasbourg. It was snowing heavily when we arrived the site. Facing a rain ditch and some big trees, I thought we had reached the boundary of the site. But the accompanying person said we were in the middle of the site. I then asked why the woods hadn't shown in the original plan. Although the trees might not be valuable, I thought that keeping them would make positive sense for the layout of the project -- at least they shouldn't be just cancelled. Next morning, I mentioned it to the owner of that French firm, and he replied the planners of this area said the trees could be removed. I said that we architects should make proactive judgments on whether to preserve these trees. We didn't participate in this project, and the reason why I mentioned it today is that we often consider European architects are mature and rational, but in fact, sometimes they can also be willful. Some architects' response to the environment is to plainly meet the planning conditions. But for me, finding even minute differences on site contributes to making the building unique, and this is the goal Land-based Rationalism pursues. As for the scheme of the consulate, I think it may be okay when it's completed, but it has nothing to do with the remains of the site. I hope our design derives from the diverse characteristics of the land. I must carefully deliberate about any changes, first and foremost, with a revered attitude, to preserve the differences. This is a matter of stance and values.

Surveying the site is only the beginning of the design, and after identifying the main design points, we shall proceed to consider how to solve problems at other levels. During the design deepening process, there will certainly be some issues that the design team need to discuss repeatedly. Yet my habit is to pin down the main design points and sketch them out right away, and thus the subsequent deepening is mostly to deal with a series of technical, functional, energy-saving and environmental protection issues within this main spatial concept. Of course, it's necessary to take into account the ideas of property owners, such as the business model or operation model proposed by the owners. We need to understand their major concerns, and how to make the space structure reasonably fit their ideas. These are secondary contradictions and should not be judged as primacy. When I handle the primacies, I have a clear sequential order in mind and will not waver easily. In no cases shall I come up with an idea on site, which doesn't match, or even conflict with the owners' requirements, and lead to failure ultimately. I rarely encounter such situations that the owners completely do not accept our design ideas. It is related to experience. I know what requirements the owners will have for a certain type of building; my accumulated experience keeps me aware of the aspects the leadership pays special attention to.

LIU Aihua:

You lead a very efficient team of architects who have produced a lot of high-quality works, and meanwhile you guide other teams of CADG through a range of design projects. What kind of methods do you use to guarantee the high quality of all these projects?

CUI Kai:

My quick comprehension of design projects comes from years of experience accumulation, which has become a habit, or a sense of excitement. When I see a case, how to identify problems and provide optimization suggestions has become an enjoyable thing for me. I am willing to do it, so there isn't much pressure.

Personally, thanks to my effective method, or you can say, strategy, I can quickly and accurately grasp the key points, knowing what is the prime strategy for certain project in certain environment, before moving on to sub-level tactics. If the initial choice is not right, which needs to be balanced or compensated through subsequent design efforts, it is by no means, as I believe, effective method. I prefer to make the right choice in the overall direction primarily and then to continue with the right choice at sub-levels.

So when I provide guidance, I will make a clear judgment on whether the major choices are correct. If yes, I will discuss with the team about the content of the next level. In most cases I play a principal role in determining the preconditions, and it makes me feel particularly accomplished when my guidance enables the design to be correctly oriented and positioned.

The opinions I provide to the team are instructive and creative, rather than restrictive. While the planning administration raises restrictive requirements, merely pointing out what does not comply with the regulations, without applicable advice, I know the limiting conditions, and can suggest where the opportunity lies. Although sometimes my opinions are practical, generally I am more willing to give inspirations on what they haven't paid attention to. I would remind architects if they have overlooked a certain aspect, or suggest that they should think more about a certain aspect. It's my duty and worthiness as the chief architect to guide and remind on the overall direction.

If the architects can understand, they will find their own solution. I mean to do so, but more often than not I can tell if they can share the same excited feelings with me. Even the architects who have worked with me for many years still have differences in excitement or quick response other than me. Therefore, for urgent projects, I will quickly draw sketches, and for less urgent ones, I will put them in my mind, and let the fellow architects keep work on the design by themselves for a week before I take the next step. Perhaps they will find a better solution, or it may not be as good as the one I figured out a week ago.

I often say that I design with values, so there may be better ways beyond the specific methods I propose. I don't think young architects need to concentrate on intimating the forms, such as how to draw sketches. More importantly, they should understand what values I mean. If there is a better solution, my sketches can be modified. If the team doesn't understand and mistakenly believe that values are just rhetoric, they can then only passively accept my ideas and take the implementation of my sketches as their priority. In reality, what I say is more important than what I draw.

In the long run, I particularly hope to explain these principles to the teams via giving guidance to their designs. If the teams acquire insights and recognition, and consequently form their own design methods based on the value system of indigenous design, they will be able to make correct judgments and choices themselves, rather than always relying on my judgment. If the teams fail to establish their own logic, they often become confused and less composed facing the requirements of the clients and the ideas of leaders. In such cases, they actually struggle with problems of secondary level, while problems of primary level have not yet been solved. If the design is only based on certain assignment book or ideas of a leader, it is passive. The proactive approach refers to our understanding of the project, and the judgments we make on behalf of future users and citizens. This is an important answer based on values. At the Inside-out School and annual excellent works selection of CADG, I always talk about some judgmental issues related to values.

After being elected as an academician, owing to this honor granted by the country, my remarks have gained more attention. But I know it doesn't mean that because of this, all my remarks should be right. The reason why a large part of my judgments are generally considered correct comes from my years of practical accumulation and research. I am clear about the right and wrong of the design, so I can keep unperturbed and will not deliberately cater to anyone. I have talked with the managers of superior supervision department about land-based rationalism and they all agree with my ideas, considering it quite aligned with the instructions given by the central government in the field of construction. As for the concepts of "cultural confidence" and "lucid waters and lush mountains", I hope the design team will not passively accept them, but will naturally use design logic and execution strategies to achieve them in specific practice, and actively envision a better living environment.

We may also encounter some difficult situations, such as inconsistent opinions from different leaders rendering the team trapped in a dilemma and unsure of what to do. At this point, I reiterate we should think again how we initially view the site, the mission, and clients' requirements, and stick to the "original intention". Some scientists turn their curiosity into a driving force for research. When it comes to design, it means keeping to the initial judgment, without being childish or completely ungrounded. To maintain a relatively mature "original intention" does require, it requires exploration through continuous trial, success or failure. The accumulation of experience is not a palpable matter that can be taught, as everyone need to do the refinement by his/her own. However, how to quickly accumulate positive experiences and reduce defeat can follow certain guidance.

I think architects should be idealists of an era, not pessimists. Some designers become pessimists when their ideals are always unfulfilled. In these cases, we can conversely analyze whether the unfulfilled ideals are suitable for this era? It is important to stay ahead of the times, to be the frontrunner. I am myself an idealist in design. Although impossible to be fully achieved every time, my ideals are realized increasingly because they are not just expressed in the language of personal aesthetics, but rather in how to better handle the relationship with the environment, the society, and the city. Thus, even if the construction quality of some projects is slightly poor, I still feel that the ideal has been achieved with a large stride forward.

LIU Aihua:
Many architects who have worked with you admired your extraordinary energy and quick judging ability. How do these merits contribute to your work methods?

CUI Kai:
I don't think I have any special innate advantages, and I'd like to take the opportunity to talk about it. In fact, I was not particularly intelligent in my childhood. Although I studied hard and wasn't naughty in school, I did not exceed teachers' expectations or have more proactive thinking. In fact, I relatively belonged to the type who was particularly afraid of teachers' questions and would duck fast when asked. I'm actually more passive and follow the crowd. I enjoyed playing with mature older children more than with those of the same age, and I was willing to learn and imitate. I was not an avid reader, either. Until now I admire people with good memory and profound knowledge. As a child, I read a lot of books, but it was difficult for me to accurately retell the content of the books. Later on, when I studied painting and copied master works, my copies were mostly less authentic. There was a gap, but I couldn't tell where it was. It was still a reflection of deficient perception to discover problems. Now I read extensively, but basically the books leave me a brief impression. I seldom use citation when writing articles. Only when the reading has left an impact on me and formed my own opinions in my mind, can I write smoothly and without hesitation. These are real experiences of my personal growth, and I am not the kind of person who has the strongest brain that never forgets.

Since learning architecture, I have gradually turned to the route of positive thinking and have benefited from it greatly. When I first took my job, good as I might be, I was not outstanding among the peers. We were all inexperienced. In the following years of work, some fellow architects gradually became frustrated with various setbacks, constrained to old design methods they learned in college without constantly observing and keeping up with the times. By contrast I am always in a state of learning and absorbing experience on my job. At beginning, I was also thwarted and might struggle for a scheme, but as I worked on, it was smoother and clearer. Whenever I make a right judgment, I got a sense of achievement, which makes me feel very happy. Thus I am always excited to receive design assignments, without feeling afraid at all. If one is afraid, the mind will become easily confused with different conditions and opinions from all parties of the project, and unable to come up with design solutions. For me, after observing relevant clues, I quickly weave them up in my mind, knowing on a whole how to solve the problem. My excitement and joy come from it. I am not as happy, nor as good at doing things other than design. Talking of my forte in design, I think I have the intensity in my creations, which is kind of interesting. I remember once I went to survey the site with two young architects. We discussed how to protect the mountain, how to adjust the city axis and the viewing site for river scenery all the way back. When we arrived at the hotel, I asked them to draw sketches, but neither of them do. Well, the problems were found without solutions, and they were afraid to draw sketches. So I picked up my pen and drew out what's in my mind. The discussions along the way had prompted me to form a strategy in a largely clear direction, not just about ideas, but rather the logic of design. When my sketches were drawn, the general design path was set.

I happened to see an architect said "the uncontroversial viewpoint is worthless". I don't quite agree with this statement. If solutions to a certain problem arouse a lot of controversy, I won't choose to follow in the wake even though some of the part may be creative. I usually prefer to look for a solution that leads to a broad consensus without too many disputes, which is especially important in my judgmental value. Sometimes my works received moderate feedback on the internet after completion, without much praise or negative evaluation. But after a long time, when suddenly someone came to tell me that he accidentally passed by a building that I designed, and he thought it was really good, I'd be very happy. I think this is the state that I need. I don't want to design an internet celebrity building that is recklessly surprising. My judgment has always been based on the common sense of the mass. If you want to produce some peculiarly great or creative works of art, my approach is not the premier choice, since it cannot provoke controversy, or attract much attention. But I think doing the right thing is more important for architectural design.

LIU Aihua:

For most architects, relevant practice experience is a requisite for being professional when facing a certain type of building. You are renowned for a multitude of diverse architectural works. How can you break free from the dependence on typical experience, grasp the core issues, and switch over different building types, to make high-quality designs?

CUI Kai:

The design I have done involves various types of buildings, including airport and hospital, which I was rarely engaged in the past. In the summary of Land-based Rationalism design methods, I particularly emphasized one point - "life oriented". When the teachers of Beijing University of Civil Engineering and Architecture helped me to sort out my lecture notes, they raised their question on this item, doubting how it would be an orientation. I believe that, the changes in life do indeed lead to the improvement of building functions and the expansion of space. It is a particularly important driving force: life is the orientation, and it means that architecture is for serving and creating a better life. Today, we witness that many inherent functions are developed to be more open, integrated, mixed, and enjoyable. The basic logic and behavioral patterns of functions are not changed, but they are no longer limited to their own. Now the main function can develop additional sections, and it has become a method of functional expansion.

Take library for example. Tables were conventionally set up for readers to sit face to face. Later on, it was discovered that a more enjoyable way of reading would be for one to sit, not towards others, but towards the scenery. It may not even be necessary to sit by the table, but rather sit on steps or lie on sofa. The way of reading is becoming more and more comfortable and personalized, as sitting at a desk cannot be said to be the only beneficial way for reading. People usually have more chance to be enlightened in their most comfortable postures, such as lying down to read books.

When people want more comfortable and personalized lives, architects should follow up. Although functions are familiar to me, we don't have to be constrained by past definitions of functions. Instead, we need to focus on people's behavioral patterns and the needs for functional expansion.

I am sensible to the innovations triggered by these functional expansions, especially during visits. For example, I visited two hospitals in Singapore and was impressed by the humanized design. Not only patients, but also doctors and patients' families were all carefully considered. Apart from medical treatment, how to arrive at the hospital, how to wait inside, and how to leave, all these aspects were given targeted considerations based on the characteristics of the hospital users.

The scale of the Singapore hospital was not large, but the details were very touching, and all design and management were user-friendly and open. For example, the parking and picking up area at the hospital entrance was designed to allow patients to sit and wait for the car, and there were wheelchairs and umbrellas placed nearby for easy borrowing. There were different kinds of drugstores, i.e. OTC drugs could be sold in convenient stores for easy access, while prescription drugs were sold in pharmacies. There were also restaurants such as McDonald's in the hospital, with the menu indicating the calorie intake. The courtyards and children's centers next to restaurants provided children with a fine place to do homework after meals. We usually think of hospitals as environments polluted by bacteria, but when greenery, courtyards, and sanitary facilities are well set, hospitals become communities where children can complete their homework, visit their hospitalized grandmother, and then go home. It's beyond imagination of domestic hospitals. The transparent laboratory left a deep impression on me as well, which made it possible for patients to watch their samples tested. The hospital required open medical area to patients so as to establish a sense of trust, just like when we see an open kitchen, we know what the hygiene conditions are and whether the working staff are earnest and skilled. Some outpatient operating rooms had windows for people to see through. This is the philosophy of Singapore hospitals, attaching particular importance to transparency and openness.

Architects should immerse themselves in the context of practical use, rather than get readily satisfied with fulfilling the clean and dirty separation of medical process and understanding its techniques in hospital. In fact, I largely rely on the experience of professional team in technique aspect; whereas my contribution to the project is to determine how to incorporate humanization and regionalism into the design.

The Kunshan Western Medical Center was my first hospital design project. At first, I was invited to participate in the project because the owner was not satisfied with the façade of the original plan. They asked me to provide some guidance. I found the project similar to those unattractive hospital buildings across the country. My suggestion was not confined to simply change the façade, but also to think about how to incorporate Jiangnan garden culture into the design, and create an environment in keeping with traditional Chinese medicine culture. Typical Jiangnan gardens are all small buildings companied with small greenery spaces. It is difficult to express the atmosphere on a long and intact façade. Therefore, I proposed to divide the outpatient building into sections, so as to make space for courtyards. At the same time, I added a long corridor for patients to rest, and to queue up.

In addition, I particularly oppose to the general practice of setting parking lot in front of the hospital. There should be a garden at that place. It happened that this hospital was going to build a large underground garage, and a subway to the hospital was under construction. Thus, we paid much attention to the underground space and the flow lines connecting the subway and parking lot. With cars all moved to underground, the front was left for a landscape yard.

Meanwhile, the waiting area was comprised a combination of internal and external corridors with seats. People can sit here to wait for medical treatment, or for surgical and laboratory results. Kunshan Western Medical Center is a large traditional Chinese medicine hospital as well as a sanatorium. Many patients are with chronic diseases. They come from nearby cities, and stay in the hospital for a relatively long time. Their families come to accompany them, too. Hence, we pondered how to provide suitable places for both patients and their families? I thought the designs should focus on the need of patients for space, to give them care and relief in their illness. In the inpatient building, we gave each bed a window, each floor a hall. There are balconies with sunshades, and a rooftop garden for leisure.

This series of considerations for practical use were combined with Jiangnan garden culture, rendering the hospital localized, which was rare in hospital design in the past. When the hospital was put into use, the hospital dean was one of the most satisfied. This was also a valuable experience for me.

The airport design is in the same vein. My first involvement in airport design was of Xiamen Airport. Mr. Yang Jinpeng, another architect in CADG, won the bidding, and he asked me to help optimize the design. From then on, I tend to pay more attention to airports when I am on business trips. Later, we were commissioned to design Jinan Airport. We have done a lot of innovations. The client said by then that although Mr. Cui didn't design many airports, he took many planes and designed from the perspective of a passenger. It was indeed true with me. I carefully observe every airport I've been to. Some airports spend tremendously on rooftops, which look great, but when it rains heavily, only a small part of the drop-off area is sheltered, and the majority have to line up and wait in their car. This design is very problematic.

In the design of Jinan Airport, I didn't attach much significance to appearance design, but gave priority to enhance the quality and function of drop-off area, making it more comfortable. Nowadays, many airports have had their roofs lifted high, which nonetheless still failed to solve the problem of shading off the rain. Later, a popular method was to add an awning underneath, as Xiamen Airport did. The problem was not solved architecturally, but operationally. It's not a good design. Therefore, when designing Jinan Airport, I strongly recommended to start with this large rooftop first, which should not only cover the rain, but also allow people to feel the sunshine and view the scenery. At night, its lighting could be gorgeous. Then we analyzed the spatial focal points of the hall, where could be low and where should be high. Many large airports were designed massively tall, which was actually unnecessary. Lower space sometimes would give people a sense of intimacy, while higher space could enhance the spatial impression of the place. With the change of space, energy-efficiency was also taken into account.

I have always believed that designs can be right or wrong, not just that many problems are ignored because everyone does it. For example, how passengers can view airplanes from inside a terminal building is not a big deal to many people, since they will see airplanes at boarding gates anyway. But I think passengers should be able to see airplanes soon after entering the waiting hall. I like T3 building of the Capital Airport. Every time I go there to take a flight, through the clear transparent glass at the very end of a narrow lane, I can see planes at the back. From there, you feel you are in the environment of an airport, rather than queuing in a large closed place. To enable passengers to quickly see airplanes, we need to make proper choices in the internal space layout, like where cannot be used for offices. Passenger flow should be aligned with the visual line, and not be obstructed. Nowadays, many small airports do not have a huge passenger flow, but the counters there are arranged in a long, long straight row with only one or two working staff. Passengers have to find a staff for check-in, and then go through security before they can finally see airplanes. This experience is actually no good.

Regarding the setting up of airport catering areas, the most common practice is to place them on both sides of pedestrian walkways like street shops. Convenient as they are, this kind of catering areas carries no scenic value. When dining at the airport, though sometimes we are in a hurry and have to leave after a bite, sometimes we may wait for the plane and sit there for a longer while. Therefore, we shall consider how to establish a visual connection between the catering area and the airside.

Many considerations from the perspective of passengers have driven us to make improvements from the overall functional layout to detailed usage with user-friendly approaches. This should be a professional responsibility of architects. As long as we think about people's experience of architectural use, buildings of any type lead to almost the same judgment, but when combined with different functions, there appear differences. The most effective way enabling us to tackle different types of projects and find design highlights is to identify the basic human behavioral characteristics behind those functional processes, to reflect on some seemingly common practices and look for improvements, and thus to solve problems.

For example, it is currently common practice that clinics in hospitals have no windows. I remember when I was a child, clinics in hospitals all had windows. When patients left, the doctors could have a cup of tea by the window. However, now the design of large hospitals is mostly flat, impossible to open so many windows. Doctors are obliged to work in enclosed space for long periods of time and with high intensity, which is really inhuman. Since the separation of the clean and polluted and the air conditioning are settled, many people think it doesn't matter to eliminate windows. In the deepening design of Kunshan

Western Medical Center, I required my team to conquer this unsolved problem, to take care of not only the patients' feelings, but also the doctors'. The deputy mayor in charge of medication and healthcare once asked me: "Patients go to the hospital for a short period of time, while doctors stay in the hospital for a long time. What are the working conditions for doctors? Where do they have meals and change clothes?" So I shared with him our considerations in this regard. By then the doctor's cafeteria was still an enclosed space, and we made special modifications to follow the mayor's advice. Doctors came to work in the early morning, and when they had breakfast, they should be able to see the sunlight and scene outside, that would make them feel comfortable. They were not supposed to dine in a closed space and leave hastily.

These adjustments are out of a comprehensive consideration of humanity, which is definitely not a slogan. I think for designers, if they propose a slogan, it must be serious and executable. They must not fall into the habit of saying empty words. If they are used to empty talks, they are not serious designers. You can't preach your ideas in this way, and end up doing things in another way. If so, you fail to combine ideas with design.

LIU Aihua:

In commissioned projects, there is often a contradiction between the inherent ideas of the owner and innovation. Innovation comes from both the architect's innovative values and the owner's expectations. How do you understand and resolve this contradiction?

CUI Kai:

If the owners have their own innovative needs, architects should fully understand them, rather than looking for other innovative directions themselves. I listen carefully to the owners' demands every time, instead of passively accepting. Even when many of the owners' opinions are not quite right, based more on economic, cost, or self-interest considerations, I will focus on the positive factor and see if I can do better in this direction. It is my habit to follow the owners' thinking and move forward, rather than thinking in another direction when dealing with such problems. In this case, the problem we need to solve is in the same direction as that of owners'. When we come up with a better solution, owners will feel what they get is worth, or surpasses what they've paid. Many times, after listening to our report, the owners have exclaimed our professionalism, implying that the plan we provided was much better than their expectation. Don't ignore the opinions of owners, or always think about persuading them change their mind. Help them to think better.

Helping owners to think better not only refers to meeting the specific interests and needs of owners, but also includes how to introduce and integrate public interests into the design. This requires careful handling: what are the potential public benefits the project may bring? How can we meet common interests without impairing the basic interests of owners? As far as a project is concerned, when the owners represent the interests of investors, the government, as an important administrative party, cares about the interests of the city and the public. Therefore, we attach much importance to the consideration and judgment of public interests in the design, which is seldom concerned by the owners. They will be more cautious in repeatedly refining the assignment book, as long as the planning conditions are met. But when we come up with a solution that not only addresses their demands, but also takes into account the overall interests of the city, the owners will feel relieved and have a sense of achievement: the project is more likely to be approved by the government, the investors' interests are guaranteed, and a balance is achieved with the public interest. We rarely make modifications again and again in the process to optimize the scheme. If we merely try to meet the interests of the owners, the government and planning administration disagree, and then if we modify purely according to government requirements, the owners cry for their interests. Sometimes architects may encounter this "drawer pulling" situation. After modified back and forth, the original plan may be completely changed in the end, just out of control. It is important for us to proactively align these two interests, and as a result, we often receive endorse from owners and planning administration departments.

LIU Aihua:

In practical projects, how do you handle other professional issues such as structure, equipment technology, landscape, interior design and so on?

CUI Kai:

A mature architect should be able to handle all the aspects in architecture, to establish positive and interactive cooperation relation with engineers of various fields so as to effectively manage and control technical scheme, and even to promote technical innovation of different fields in their application for architectural design. This is my understanding over the years.

At first, my understanding in this aspect was just about how to make these professional fields well serve architectural design. Of course, the first step is to understand the basic principles and rules of these technical designs, the forms and characteristics of equipment system. I learned a lot about these fairly early. Not long after I began to work, I did pipeline overall design and drew some technical drawings. The knowledge lingered in my mind for a long time. I clearly knew how to arrange pipelines more reasonably. It is very important to grasp the knowledge and turn it into the intellectual base for you to discuss problems with equipment and structural engineers. Otherwise, it will be subject to division of engineers, and that will be relatively passive for us. Some young architects lack experience of this kind. So when I guide design, it's not entirely about guiding concepts. Many a time it's about correcting design errors.

In deepening the design of Xi'an Beilin Museum, the design team worked hard and spent much time on the design. However, when they came to me for consultation the other day, I found some major problems were still unsolved. One side of the main street that were designed for stores were now all occupied by air ducts, with walls covered with louver gratings everywhere. I said we should have discussed these matters earlier. Was it appropriate to place the air ductwork here? Although the client seemed not to care much at this time, we should consider preparing premises for future business operation, which was for space use on city dimension. Not because the air ducts were there, should the stores give way. We needed to find a solution, such as whether the air duct could be perpendicular to the street. After I put forward these suggestions, the design team really solved the problem. I also noticed a pillar was in the middle of the parking lane. I checked several floor plans and found that the pillar could be pulled out, and it subsequently turned out that I was right. Then why no consideration was given earlier to pull it out? As for the design of restroom, we should think over how to make use of the space more efficiently, and how to screen the view at the entrance. More often than not I also place much attention to these details.

A mature architect must have this kind of "mysophobia". Details actually indicate the quality of architectural design. Architects must be able to think proactively for clients and minimize errors in design. These are the basic merits a mature and excellent architect possesses. What's more, we are now working on green buildings, and should take more initiatives to guide relevant professionals to adapt to this trend, focusing on the requirements of low-carbon and energy-efficiency for structure and equipment technology. I think this involves a series of innovations. I don't consider innovation is just about conceptual or formal matter, it should be a series of internal innovations on technology and quality.

The industry now applies a not-so-good system, which is to bid again for subsequent engineering design after the plan is made. The deepening design and construction drawing design are hence often done by the EPC team of the general contractor. Whenever I talk about this with the owner, I try to explain in every detail that the innovation and high-quality completion of the design rely on a series of proactive responses and adjustments from the phase of scheme design to construction drawings design and even to construction cooperation. Should it be set apart, the later part of the task will be most unlikely to be completed with high quality. The constructor may only be responsible for the appearance of the scheme, while the interior is adjusted and pieced together at will for the mere sake of sparing money and effort. Since the constructor is not required to contribute creativity when taking over the task, it will assign other architects to follow up, just making sure the standards are met without problems, and the work will pass review. When interior quality control is not guaranteed, decoration is the solution for all, let alone technological innovation, of which many opportunities have been missed.

Therefore I always emphasize the whole process innovation and high-quality design from beginning to end. It is of particular importance. It means that there must be continuous management and guide for other professional aspects, including interior design, landscape design, lighting design, and curtain wall design, which are now revised once or more for almost every project. Mr. Wang Dongning, who has been cooperating with us all these years, is an excellent lighting designer and has won many international lighting awards. He has felt it rewarding every time he listens to my opinions. The landscape design and so on are the same. My analysis on the background of project and my thoughts and basic values help to reduce the detours in respective designs, and keep them from being hidebound by routines at the beginning, which will otherwise lead to disconnection from the overall design concept.

Architects not only need to have a sense of active participation in the design of various professional aspects they collaborate with, but also actively implement the architectural concepts in the process of their participation. I understand sometimes it's not feasible for architects to implement all design ideas, for reasons caused by the owner, the building cost, or the constructor. Nonetheless, there should be no major mistakes. For example, if we only guide the structure and equipment design in line with architecture design, we achieve 80 points. If we go on to supervise indoor and landscape design, we may achieve 90 points. If we do a good job with curtain walls, we may achieve 95 points. However, if we do not do well with door handles and staircase handrails, we may lose the remaining 5 points. Still, it is important to try to reach as far as possible for every project. Basically for simple projects, we can score 99 points, while for others we may score a bit lower. But if the entire system is well controlled, we can probably score above 90 points. So I think architects, as design principals, should have idealism and strive to get closer to 100 points. This is only about design and construction. Whether the appropriate usage purpose can be achieved is also very important. For example, although the Jiangsu Garden Expo is built during the pandemic, it's operated under very well. Just this morning, architect Ms. Dong Yuanzheng told that the owner said the multi-functional hall in the hotel was not big enough to hold large-scale weddings. They suggested to expand it. We are certainly happy to continue with this cooperation. Projects in particularly good usage state like this can score up to 110 or 120 points.

As I can see more, observe more, remember more, I've become increasingly mature, and consequently, I am able to turn design into a universal one regardless of intra-professions division. We can work in other fields like graphic design and logo design. Recently, owner of Quanzhou Railway Station wanted to erect a sculpture and asked if I could do it. At first I rejected, and architect Mr. Ren Zuhua of my team planned to find a friend from an academy of fine arts to do it. Afterwards, it occurred to me that the work shouldn't be entrusted to ordinary artists unfamiliar with the local environment. Hence, I drew a sketch myself. To display the beauty of ridges and tiles on the large roofs of Fujian residential buildings. So was it sculptured accordingly later on, and the client liked it very much. I dare not say how highly accomplished it was from an artistic perspective, but it did come from my knowledge and understanding of local culture. What's expected for the square of a railway station should be something able to showcase local characteristics, rather than a purely artistic modeling.

LIU Aihua:

You have participated in design competitions and bid evaluations for many major projects. As a judge, how did you evaluate different entries?

CUI Kai:

Frankly speaking, the competition is a test not only for design teams but also for judges. I don't think the judges have the right to make willful and self-centered judgments, nor can they review any proposals at will. Therefore, in the first place we should respect the efforts of design teams, and respect the expectations of owners for the construction of the project. It is the basic idea I bear in mind in every review that we will obtain a relatively good solution through our evaluation. Thanks to my efficient and extensive design practice and experience through day-to-day work, I am able to quickly identify problems, or in other words, think of problems that need to be solved. From this perspective, it's a bit like reviewing with empathy. If I were the designer, what should I do to design this project, instead of acting like an outsider, using general design requirements as a measuring gauge?

I attach much importance to the environmental analysis and final design report of the scheme, while the body part of concept statement is less cared. If time is limited, this part will be roughly passed because many concepts have been common to everyone. The important thing is how to implement the concept into design, which actually depends on the logical relationship between the site analysis and completed design. The judges are not moved by general or conceptual things, but rather focus more on design works. I have an innate preference for designs completed with high quality. Even if there are still issues or even controversies remain unsolved in the scheme, I believe it is worth being taken seriously. Some design entries perform poorly in their completion of expression, making it difficult to truly create good designs even if there are certain points accurately marked.

Meanwhile, in many evaluations, especially those in terms of urban designs, values are always involved. For example, when the design aims to solve problems related to a small or medium-sized city, but the design team takes successful experiences of world-renowned metropolises as reference to illustrate the design concept, it is complete detachment from reality. So I pay more attention to the problems that the city urgently needs to solve, to its potential for development, and to the kind of composure or confidence it should exhibit, while I strongly object to deceptive design. Although a design starts from competition, the stance and approaches of the design team are actually important judgment points.

Next we review the comprehensive expression quality and intention of the completed designs, whether they appropriately reflect the needs of the assignment without exaggeration or excessiveness. Apart from the appearance design, internal spaces should also be reviewed. Teams lack good skills usually focus on the appearance, but when their layout and space are examined, there is nothing. This kind is the first I choose to reject. If the design appears to contain bit of exaggeration, which however brings a very good spatial framework, I think it is also commendable. I firmly oppose the design having only the appearance. Not only should it include more than appearance, but the space should also reflect the cultural characteristics of the region, with innovative guidance in functions, and considerations for public interests. For example, for a building design, when it takes into account the openness and relationship with the surrounding urban environment, it will be very positive and valuable.

Overall, I make evaluation on these levels: firstly, whether the problems are accurately analyzed; secondly, whether the team delves into the design sufficiently; thirdly, whether the design focuses just on the appearance of building; fourthly, whether the design presents the values of design team.

I make quick judgments about the above aspects when reviewing the entries. We also listen to the presentation of design teams. When I listen, I take some notes, but I still need to look at the drawings. For schemes that are not good enough, I will not take a second look; for those that are all right, I will examine more carefully; and when I come across good schemes, I will try my best to read them carefully and pick out foibles. Hence, I often write opinions with design optimization suggestions in the review comments, rather than simply remarking which is good or which is not. For the selected design entries or those evaluated among the top, I am particularly willing to affirm their values. My positive comments even become supportive quotes for the clients to report to their superior leaders. There have been several times when I was asked to report to the owners on behalf of the judge committee, because I was recognized not only being able to write comments well, but also present well.

I think it is important for judges to have a sense of empathy. This does not mean that all the judges must be architects with practical working experience. But if the judges also do designs themselves, they will put themselves in certain position and think about the solution for certain problems. I participated in many school design competitions in Shenzhen, and there were several younger architects among judges at that time, like Mr. Dong Gong and Mr. Meng Yan. They reviewed more closely, talked more carefully, and hit the nail on the head. So I think it is an important guarantee for a project to have excellent designers as reviewing judges, as it can promote good designs and block out inferior ones. In case the judges are less percipient, and if they seek personal or related interests, they are likely to make poor judgments. Therefore the reviewing judges are very important. I am sometimes reluctant to participate in certain design competitions because I may worry about the competency of such judges. They don't care about the design or the assessment of the site.

LIU Aihua:

In this age of information explosion, facing the massive number of books, magazines, and online news, how do you choose to read and connect them with design practice?

CUI Kai:

Reading is not the most important way for me to obtain architectural design information. Sometimes I do need to refer to books, but it is not a systematic and planned type of reading, and I seldom get online for information. However, I attach great importance to personal experience. Over the years, I have had many opportunities to conduct research at home and abroad, which has greatly benefited me. This year, with the help of my wife, I systematically sorted the photos of buildings I have studied over the years, which is now very convenient for reference. Recently when preparing courses for Beijing Construction University recently, I also selected cases of foreign buildings that I had visited, so that I could explain them in detail and depth. Take the medical building we just talked about for example. I visited Khoo Teck Puat Hospital in Singapore in 2012. Many of its details had influences on the design of Kunshan Western Medical Center.

Hence, in the face of a large amount of data, the important thing is how to make use of it in design. Every time I look at the photos, I feel as if I returned back to the scenes when the photos were taken, and this is a different reading experience for me. So I think highly of personal experience, which is a very important way for me to improve my design.

I also browse major architectural publications such as Architectural Journal, Time Architecture, and World Architecture. As if looking at reference, seeing excellent works from peers is likewise a spur for me. Sometimes the works are very good on the whole, but there are certain areas that I personally feel are too affected and overly designed. When I look at them, I make my own judgments about which areas can be solved with more appropriate methods and scales.

So personal experience and memory, as well as reading with my own understanding and viewpoints, are two important learning methods for me.

Recently, I have been listening to records of Mr. Yu Qiuyu's lectures for two months, and I listen to them every day even when I am walking. I think what's admirable in his lectures are the descriptions of many scenes of Chinese culture, which have stricken a particular chord with me. It is not a kind of knowledge-based learning that I should go from A to Z, but rather a learning that arouses sympathies. It makes me feel that my thinking, my work, and my judgment of values are all in the large context of Chinese culture, which makes me increasingly assured that I live and work clear-mindedly.

LIU Aihua:

In every quarterly exhibition in your studio, you write a poem for the exhibited project. Last time I saw on your sketch book, there was also a poem titled Towards the North written next to the conceptual sketch of the Southern Song Dynasty Imperial Mausoleum Site Museum. Why do you choose to use poetry to express your design ideas?

CUI Kai:

Without my noticing it, I developed some literary preferences when I was young, and wrote some limericks. For me, writing poems is a habit to use concise language to summarize complex things. Of course, it is based on sentiments, and without sentiments, I cannot write poems. The refinement of language comes from the judgment of design. Regarding why I wrote Towards the North, when I was doing that design, I browsed through books related to the Southern

Song Dynasty; it happened that a classmate of mine once gave me a book on the study of urban pattern of the Song Dynasty, but I left it unread; that day, I flipped through it and saw that most of the buildings in Lin'an during the Southern Song Dynasty, including the city gate, had their main entrances facing towards the north; this inclination of direction caught my attention, because we were too accustomed to buildings facing south, and we subjectively thought everything was in the south. But when it came to the Southern Song Dynasty – I suddenly realized that it was actually a nostalgic feeling that they wanted to return to their northern hometown. I think this was the emotion that needed to be expressed in design, so it took me almost no time to compose in the linguistic form of poem Towards the North, expressing sentiments as well as the organization of my design. This poem was a part of my design meditation.

The project was near Shaoxing. When I went to survey the site, I walked through tea fields under the scorching sun all the way. The accompanying person introduced that there was nothing on the site, but during the Southern Song Dynasty, a beautiful house was built on top of every tomb of Emperor, and the coffins were carried in by boat from Lin'an to the mountain pass. I felt it very interesting, quite different from the Eastern and Western Tombs of the Qing Dynasty, which I surveyed in college, and the Ming Dynasty Tombs. As a rule, as soon as the emperor took the throne, he began to search for a place for his cemetery and then had his tomb built there for the rest of his life. It was the territory of the monarch and a place for them to long rest. By contrast, in the case of imperial mausoleum of the Southern Song Dynasty in Shaoxing, decreased emperors were not buried after death. When I told the client about these sentiments and the consistent design ideas, they were also much interested and changed their initial thinking that there's nothing here and they had no ideas how to build a museum.

I was not born a lover of history. Mr. Yu Qiuyu's lectures on the culture of different historic periods have benefited me a lot. The Wei, Jin, Northern and Southern dynasties were a prime time for culture. Although there were many poets in the Tang Dynasty, few could surpass Li Bai and Du Fu. Tao Yuanming was a representative figure in Jin Dynasty. We happened to be working on a cultural complex in Youyang, Chongqing. Youyang is said to be the proto-site of Peach Blossom Land in Yao Yuanming's article, where there are huge karst caves, beautiful countryside, and many corresponding historical records. So when I did the design, associating with Tao Yuanming's seclusive and noble literati sentiment, I had a feeling that I was communicating with him from afar by design.

In the past, the expression of architectural design on cultural heritage was often passive, largely confined to Chinese glazed tiles and big roofs which were quite formal but constrained to a specific period of time. They reminded people of those who tried to restore the Qing emperor during the Republic of China. We couldn't comprehend, nor had any interest in these archaized buildings when we were young. I think buildings like the National Art Museum of China designed by Mr. Dai Nianci, and the Cultural Palace of Nationalities designed by Mr. Zhang Bo, have reached the peak of perfection in this direction. No successors have the ability to do better.

Therefore, we should turn to the direction of how to break away from the simple pursue of form to carry out cultural inheritance. Just as the remark by Lao Zi is still a classic in Chinese architecture, and many scholars internationally know it, I think using this cross-time cultural perception to express cultural inheritance opens a broader space for us. At present, many outstanding architects have studied and applied the essence of mountains-and-waters paintings, literati's feelings, and distinctive life context in their designs. These thoughts on inheritance have jumped out of the strange loop of modernization of national forms that restrained several previous generations. They are for the pursuit of cultural resonance.

LIU Aihua:

Do you prefer to read more books in the field of architecture or books outside of your field?

CUI Kai:

I read some books in the field of architecture for once or twice, but there are not many. I have been reading more on biographies recently, and memoirs and commemorative works written by architects, which I am quite interested in. I mainly read about the history of that time, the idealism of people in that era, their thinking and confusion, so as to obtain a deeper understanding of the relationship between people and society. I think that now at my age, I'm more inclined to compare my years of work and accumulated thoughts with those of the predecessors, to know what Mr. Liang Sicheng thought, what Mr. Lu Sheng thought, how they handled the things they encountered, and what inspiration they left for future generations. I also enjoy biographies of people other than architects, and I used to read life stories of Churchill, Obama, and Steve Jobs. Recently, I have been reading the biography of Jane Jacobs, a social worker's observation, protest, and struggle of the city.

This kind of reading is a bit like looking at the historical positioning of what we are doing today. It may sound a bit big, but I believe that architects, as important builders of an era, should first have idealism, and second, have a responsible attitude towards history, so that we can have a broader vision, and our designs will contribute to make cities better and rural areas better. Although we know that what architects can do are limited, we shall still do our work with such a big aspiration, and whatever achievements we make, I think this is very important. I'm not in favor of those who do not respond to the needs of society and only mind their own business with the arguments that they cannot accomplish feats, or they are dissatisfied with the current social reality and have many dissents. This is seclusion. Among Buddhism, Taoism, and Confucianism, I personally think Confucianism goes well with me, and Buddhism is good as well. Taoism features a strong personality and cannot tolerate integration into society. When one is satisfied with self-escape, it is not good for society. Anyway the ideas of Taoism are still beneficial to society. But Lao Zi left only five thousand words in his lifetime. These five thousand words are great, but he could have left tens of thousands of words, right? After all, Li Bai left behind so many good poems.

So for me personally, I am not particularly concerned about which one of my works is the best or which thing I've done is the most meaningful, and then make light of other works. Some people may think that for some projects, we shall seize opportunities and do a great work, while for those less prominent ones, we can make them a profitable job. I think this opportunistic stance is harmful to cities and not good for architects themselves, who play a role of 'changing face'. If one takes out only good works and repeatedly talks about them, and stashes the bad ones and keeps them unannounced, even when the proportion of bad or average works is far bigger than good works, do you think he is a responsible architect? He is at best a talented and intelligent architect. My requirement for myself is to do good design for every project, to be worthy of the given opportunity and this land. It is particularly important for me. Sometimes I'm asked why I take over so many projects. My answer is I can thus make more contributions. More good architectural works mean more contributions to society. I am busy but happy, and because of this idealistic mindset, I have higher demands on myself. I try to do my design better, and in the process keep a state pursuing for progress and exploration, without being self-satisfied and resting on my laurels.

LIU Aihua:

How do you maintain a long-term sensibility to society and keenness to innovation throughout your years of work? And how do you keep breaking through established cognition, and constantly gaining energy for self-renewal?

CUI Kai:

I think continuous learning is a basic requirement for architects. This profession serves almost all industries, and most of the spaces in life need architects to design. What architects have learned at school are basic techniques that have limitations. After their graduation, some technological processes and architectural features may have undergone fundamental changes with the development of times. The continuous education for architects in foreign countries is also a professional necessity, as not all professions have this sort of industrial requirement. Continuous learning is professionally required for architects, which is the first point.

Secondly, CADG has given me such a platform. When I first came to work in the institute, and caught sight of so many famous architects, I understood my self-requirement should be different from that of going to a small design institute. Later on, working with so many excellent colleagues, at that time newly recruited competitive university graduates, I was further motivated and hoped to gain a foothold here. Therefore, I was not disturbed by external interference such as the trend of going into business.

Meanwhile, in participating in industry communications such as the activities of Architectural Society of China, I received much encouragement and care for us younger generation from the senior, which subtly became a driving force for me. Although sometimes I can't name accurately who taught me in which matter – of course I learned a lot substantially from my mentor Mr. Peng Yigang – in these academic communications, what I heard and saw was more about cultural literacy and spirit.

With the progress of the times, architectural design has been constantly innovating, especially after the year 2000 when foreign architects came into our sight and broadened our vision. That included the gaps we observed when going abroad, and the high starting point of architectural works by overseas returnees. They have more extensive knowledge, as well as theoretical and practical achievements obtained under the foreign education system, all of which make me feel that this is a highly competitive industry. I don't consider myself among the top players of my generation just because I was the first batch of college students after the Cultural Revolution. Instead, I am constantly propelled to keep up with the industrial progress. All these are the driving forces pushing me ahead.

After identifying the direction of Land-based Rationalism, I find that I have entered a more rational, confident, and clear path in creative thinking, with the works systematically improved, and the success rate raised. In this situation, I need to maintain a higher pursuit, rather than constantly repeating myself. Therefore, we have proposed some research fields for technological development. For example, we applied as-cast finish concrete in practical projects, and solved relevant technical problems via design. In another case, starting from an occasional proposal, we made successful attempt to use wood structure, and now we've created opportunities for more attempts, as wood structure is a particularly good material for green and low-carbon purpose. It's carbon sink material. Once again, in terms of lightweight structure, we used to complain about "fat beams and thick columns", which originated from the conservative design of the structure and the lack of involvement of architects. We are now combining new projects to create lightweight steel structures, or lightweight structures or light structures. Moreover, we have different thoughts and practical explorations in precast assembly, which has received much attention from the industry. We are constantly looking for goals of different phases, focusing on technology and green building innovation. These can be specifically proactive.

Sometime when I read Architectural Journal, I used to think that our works, including those of CADG, were quite good. However, the works now included in Architectural Journal are mostly impressive and worth studying, to which I'd ever want to take a visit when I have time. So I think if we only compare each other within a small circle, assessing who is doing better or whose position is higher, it's a bit meaningless and narrow-minded. I often tell young architects that we must have an industry-wide vision. It doesn't necessarily require everyone to have an international perspective, but we must take the whole industry into account. The fact is that in the same Chinese context of creation, there are a group of excellent architects making excellent practice; from creativity to mastery of advanced technology, to high completion, these are all traits that our architects of major organizations need to actively learn and cite as incentives on us. That's why I hope large design institutes provide teaching and urge everyone to write more articles and participate in various academic activities. I think it is very important to keep up and see beyond one's surroundings. This is what I particularly want to remind the young and middle-aged architects succeeding my generation, that if they have any academic and professional pursuits, they need to make higher demands on themselves. This is my sincere suggestion.

When I won the Liang Sicheng Prize, I wrote an article In the Middle, in which I described a state of mind that not everyone could understand, but was the description of my thought. The 'middle' is a very important position, which is about the attitude towards being inside or outside the academic circle, being in large or small design organizations, and being young of the old generation. There is actually a lot to learn in this "middle". It is necessary to maintain openness, neither deliberately choosing sides, nor being radical and indignant, nor blindly self-appreciating and resting on laurels. I'm often invited to give speeches. It is because by keeping open-minded, I am able to stay innovative and thus always have new things to talk about. In the transition of times, there have emerged many new problems requiring us to rethink and confront. In a word, it is vitally important to keep up with the times and constantly motivate ourselves in the development of the times.

后记

《本土设计Ⅲ》是继之前《本土设计》和《本土设计Ⅱ》出版以来对本土设计理论和方法的持续探索和实践的又一阶段性总结，算起来自2009年这个观点的提出已有十四个年头了。这十几年来也正值国家社会经济从快速发展高潮期向平稳期过渡转型的阶段，新时代所提倡的关于生态保护和文化自信的许多战略方向使本土设计的探索有了更有力的支撑。许多同行希望我对这个方向进行进一步的梳理和完善，形成可以借鉴和传播的某种模式语言，而两年前我和助手们应邀在北京建筑大学开设"本土设计概论"课也是一个很好的契机，让我们在忙碌的设计工作中更努力地研究、学习、思考，为这本书得以顺利出版创造了条件。在这里我要表达为此付出辛勤劳动的各位同事、朋友、家人的深切感谢！

首先应该感谢的是和我一起工作的本土设计研究中心的各位同伴以及中国院许多设计团队的同事们，看着这么多建成项目的照片，无不让我想起其背后大家紧张而愉快地工作的情景，我总会说这些作品是大家共同的成果；其次应该感谢我的老朋友、老同事张广源先生，他几乎亲自拍摄了我所有的作品，并常常提出一些很有观点的建议和忠告，他甚至把这本书的出版当作他真正退休前要完成的最后一项工作，让我深深地为老友的这份真情所感动！在他的带领下，任浩、刘爱华、谭雅宁、傅晓铭等同仁从前期策划、文稿整理、翻译、照片选择、排版、后期印刷制作等等方面事无巨细做了大量工作，没有他（她）们的督促和提醒，我也很难及时补充相关资料，书的顺利出版有赖于这个小团队的协作和努力！然后我要特别感谢王建国院士在百忙中为这本书亲笔作序，他以学者的视角对本土设计深入而系统的点评大大提升了本书的学术价值；另外我在书中还收录了同济大学出版社支文军社长、北京建筑大学金秋野老师和清华大学范路老师对本土设计探索的评论文章，他们的关注和评价使我们的工作被纳入到中国当代建筑学发展的大背景之中，赋予了某种使命感和历史意义，对此我和大院的同事们都心存感激；我还要特别感谢中国建筑工业出版社的徐晓飞先生，他不仅是此书的责任编辑，而且之前他就职于清华大学出版社时就是《本土设计》作品集的责编，感谢他一直以来对我们的热情付出和认真负责的态度；最后我要深深地感谢我的夫人周蔚，她不仅长久以来默默地支持我的事业，承担了全部的家务劳动，而我常年的忙碌工作也是以缺少对她的关心和陪伴为代价的，每每想起我内心都十分愧疚和感恩！

其实此刻我还十分怀念我的恩师彭一刚先生、聂兰生先生，还有近期先后仙逝的关肇邺先生、钟训正先生、栗德祥先生，他们曾经给我的指导和勉励一直是我前行的动力，我想这本书也应该是学生晚辈交给先生们的一份新的答卷。

崔愷
2023年7月7日
写于从风凉的丹麦飞往酷暑的北京途中

EPILOGUE

Land-based Rationalism III can be seen as a sort of phased summary of my exploration and practice of Land-based Rationalism theory and methodology after *Land-based Rationalism* & *Land-based Rationalism II* were published. It has been 14 years since I proposed Land-based Rationalism in 2009, and during these years China has been going through a transitional period from the peak of rapid social and economic development to a stable period, and many strategic directions in the new era regarding ecological protection and cultural self-confidence have provided stronger support for Land-based Rationalism. Many colleagues encouraged me to further improve my work in this direction to form a pattern language that can be promoted. Two years ago, my assistants and I were invited to lecture on "Introduction to Land-based Rationalism" at Beijing University of Civil Engineering and Architecture, which was a great opportunity for us to learn and research in our busy working routines, facilitating the successful publication of this book. I would like to express my heartfelt gratitude to all my colleagues, friends, and family members who have contributed to this book.

First of all, I would like to thank my colleagues at Land-based Rationalism Design & Research Center and many other colleagues from CADG. Flipping through photos of so many completed projects, I am reminded of the days when everyone was working nervously but happily, and I always say that these achievements belong to everyone. Secondly, I would like to thank Mr. ZHANG Guangyuan, an old friend and colleague of mine. He took photos of almost all of my works and often gave me insightful suggestions. He even regarded the publication of this book as the last task before his retirement, and his sincerity as an old friend deeply impressed me. Under his direction, REN Hao, LIU Aihua, TAN Yaning, FU Xiaoming, and other colleagues have made great efforts in the planning, editing, translation, photo selection, design and printing of the book. Without their efforts and feedback, it would be difficult for me to optimize the book in a timely manner, and this small team's efforts have guaranteed he smooth publication of the book. I should also express my special thanks to Academician WANG Jianguo for writing the foreword amidst a busy schedule. His in-depth and systematic comments on Land-based Rationalism from a scholar's perspective has greatly improved the academic value of this book. In addition, commentary articles on my exploration of Land-based Rationalism by President ZHI Wenjun from Tongji University Press, Professor JIN Qiuye from Beijing University of Civil Engineering and Architecture, and Professor FAN Lu from Tsinghua University are also included in the book. Thanks to their concern and evaluation, our work has been put in a broader context of contemporary architectural development in China, giving it a sense of mission and historical significance, for which my colleagues and I are very grateful. I would also like to express my special thanks to Mr. XU Xiaofei from China Architecture & Building Press, who is the editor in charge of not only this book, but also the previous *Land-based Rationalism* when he was working at Tsinghua University Press. I am thankful for his enthusiasm and conscientious attitude for the publication of the books. Finally, I would like to sincerely thank ZHOU Wei, my wife. She has been supporting my career for a long time and has taken on all the household chores. My full engagement in my career comes at the cost of lack of care and companionship for her, which often makes me feel guilty and grateful from the bottom of my heart.

At this moment, I can't help remembering Mr. PENG Yigang and Mr. NIE Lansheng, who were my mentors, as well as the recently deceased Mr. GUAN Zhaoye, Mr. ZHONG Xunzheng, and Mr. LI Dexiang. Their guidance and encouragement for me have always been the source of my strength in pursing my career. I hope this book can serve as a sort of report delivered from a student to them.

CUI Kai
Jul. 7th, 2023
Written on the flight from the coolness of Denmark
to the scorching heat of Beijing

附注
ANNOTATIONS

注1 节选自《时代建筑》2021年第4期《寻溪问巷，理水作庭——记南浔城市规划展览馆的本土营造策略》，文：邢野、崔愷。
注2 文：喻弢、金爽。
注3 节选自《建筑学报》2023年第4期《医疗建筑的人文性、本土性探索——昆山西部医疗中心的设计策略》，文：崔磊等。
注4 节选自《建筑学报》2020年第5期《文化聚落——敦煌市公共文化综合服务中心设计》，文：吴斌。
注5 节选自《建筑技艺》2023年第3期《"亦传亦奇"的纪念性——枣庄铁道游击队纪念馆设计概述》，文：邢野、高凡。
注6 节选自《建筑学报》2022年第8期《"再生"的花园——第11届江苏省园博园主展馆片区设计》，文：董元铮、崔愷。
注7 原载于《建筑学报》2018年第04期，文：崔愷。
注8 文：关飞。
注9 节选自《建筑技艺》2022年第11期《城里的"山"——崇礼中心设计与建造札记》，文：康凯、张一楠。

本书中，作者对已刊发的文章、节选文字重新做了修订。

Born in Beijing, 1957

Master Degree of Engineering under the guidance of Professor Peng Yigang, Department of Architecture, Tianjin University, 1984
Architect, Architecture Design Institute, Ministry of Construction, 1984

Architect, Hua Sen Architecture & Engineering Design Consultants Ltd., 1985
Senior Architect, Vice Chief Architect, Architecture Design Institute, Ministry of Construction
Vice President, Chief Architect, Architecture Design Institute of Ministry of Construction, from 1989 to 2000
Chief Architect, Architecture Design & Research Group, Co. Ltd. (CADG) since 2000

Deputy Board Member, UIA, from 1999 to 2008
Speech in Kuala Lumpur, Malaysia, 2001
Speech in Busan, Korea, 2004
Judge of International Competition, Busan, Korea, 2004
International Architecture Exhibition, Taipei, 2004
Vice President, Architectural Society of China, 2004
National Master of Engineering Survey and Design, 2000
French Culture & Art Cavalier Medal, 2003
Founded Cui Kai Studio, 2003

Speech in Seoul, Korea, 2005
Speech in Singapore, 2005
Biennial in Shenzhen, 2005
Speech in Colombo, Sri Lanka, 2006
Judge of Hong Kong Institute of Architects Annual Awards, 2006
Speech in Delft, Netherlands, 2006

1984 - 2000 | 2000 - 2004 | 2004 - 2006

Er Pang Gong Hyatt Hotel, Xi'an ■

No.3 Villa, Commune by Great Wall, Beijing

Yinxu Museum, Anyang ■ ■ ■

Ming Hua Center in Shekou, Shenzhen ■

Tsinghua Innovation Center, Beijing ■ ■

Desheng Up Town, Beijing ■ ■

Feng Ze Yuan Hotel, Beijing ■ ■ ■

Office Building of China Academy of Urban Planning & Design, Beijing ■

Capital Museum, Beijing ■ ■ ■ ■

Huairou Training Center, Ministry of Foreign Affairs, Beijing

Focus Place, Beijing ■ ■ ■

Lhasa Railway Station, Lhasa ■ ■ ■

Office Building of Beijing Foreign Language Teaching & Research Press, Beijing ■ ■ ■ ■

Rebuilding of ACCA Office Building, Beijing ■

No.9th Office Building of Dalian Software Park, Dalian ■

Modern Town Apartment Tower, Beijing ■

Artron Color Printing Center, Beijing ■ ■ ■

Software Engineer' Apartment of Dalian Software Park, Dalian ■

Yifu Building in Beijing Foreign Studies University, Beijing ■ ■

International Conference Center of Beijing FLTRP, Beijing ■

No.10th, 11th Office Building of Dalian Software Park, Dalian ■

Weihai CITIC Financial Building
Renovation of FLTRP Printery
Beijing Modern Town Kindergarten
Library & Catering Service Center of China University of Petroleum (East China) Dongying Campus

Renovation of Office Building of CADG
Ningbo Tianyi Homes
Beijing Kangbao Garden
Chengdu Oriental Pearl Garden
Beijing Blue Castle International Apartment
Tianjin TEDA Ploytechnic
Tianjin University of Science & Technology, TEDA

No.8th Office Building of Dalian Software Park, Dalian
Xi'an Ziwei Villaes
Public Service Area of TEDA Higher Education Park, Phase I
College of Life Sciences, Zhejiang University
Tangshan International Exhibition Center
Zhuzhou Sports Center

AWARDS   ■ National Best Project Design Award   ■ Best Design Award of the Ministry of Construction   ■ National Best Project Design Industry Award

2019年作为颁奖嘉宾出席庆祝中华人民共和国成立七十周年中国建筑学会建筑创作大奖颁奖活动，在2009-2019年中国建筑学会建筑创作大奖的100项作品中有8个作品入选 2019年参加深港城市建筑双城双年展"城市之眼"主题展 2020年为《建筑学报》撰文《大瘟疫提醒我们要思考什么？》 2020年被聘为"一带一路"建筑类大学国际联盟大学生建筑和结构设计竞赛评审特聘专家 2020年作为终审评委参加"自然建造·Architecture China Award"颁奖	2020年策划"城市的进化"展览及论坛 2021年"本土·季展"专题展开幕，计划每年分三个主题展示本土中心不同研究方向的成果 2021年接受凤凰卫视中文台《筑梦人》栏目专访 2021年主编的《绿色建筑设计导则》出版发行 2021年参加第17届威尼斯国际建筑双年展中国国家馆主题展 2021年主持的"本土设计实践导论"研究生课程在北京建筑大学开课	2021年获得国家科技进步一等奖 2021年主编《地域气候适应型绿色公共建筑设计研究丛书》出版发行 2021年作为演讲讨论嘉宾参加中国-巴西建筑论坛 2022年策划中国建筑设计研究院成立70周年主题展馆"中国建筑设计研究院院史陈列馆"开馆 2022年策划"本土设计—城市更新巡展重庆展"开展

2019 - 2020 | 2020 - 2021 | 2021 - 2022

北京世界园艺博览会中国馆 ·北京
全国绿色建筑创新奖一等奖
全国优秀工程勘察设计行业一等奖
中国建筑学会建筑设计奖公共建筑一等奖、绿色生态技术一等奖

遂宁宋瓷文化中心 ·遂宁
北京市建筑工程设计综合二等奖、绿色专项三等奖
中国建筑学会建筑设计奖公共建筑三等奖

天府农业博览园主展馆 ·成都
Wood Design&Building Awards 优秀项目奖
Structural Awards 2022最佳项目奖

京藏交流中心 ·拉萨
北京市建筑工程设计综合三等奖

荣成市少年宫 ·荣成
FX Awards-GLOBAL PROJECT类别大奖

青岛上合之珠国际博览中心 ·青岛

铁道游击队纪念馆 ·枣庄

张家港金港文化中心 ·张家港

崇礼中心 ·崇礼

宝鸡文化艺术中心 ·宝鸡
陕西省优秀工程勘察设计一等奖

江苏园博园主展馆及傲图格精选酒店 ·南京

重庆规划展览馆 ·重庆

南浔城市规划展览馆 ·南浔
北京市建筑工程设计综合三等奖、人文建筑单项二等奖

江苏园博园未来花园 ·南京

徐州园博园儿童友好中心 ·徐州

日照市科技馆 ·日照
北京市建筑工程设计（BIM）优秀奖
AEC Excellence Awards 建筑运算类优胜奖

武汉大学城市设计学院教学楼 ·武汉

湖州市滨湖高中 ·湖州

雄安站 ·雄安

昆山小桃源 ·昆山

科兴天富厂区更新改造 ·北京

威海图书馆及群艺馆
保定市关汉卿大剧院
景德镇锦荣外国语学校
成都中车共享城启动区
北京电影学院怀柔校区
北京外国语大学西校区教学服务楼群

益阳市民文化中心及政务中心
天津大学建筑学院建筑北馆
大兴机场北线高速廊坊站
延安新区全民健身运动中心体育场
榆林市东沙文体馆
舟山文化中心
昆山西部医疗中心一期
象屿集团大厦

景德镇艺术职业大学一期
清华大学深圳国际研究生院一期
中国人民大学通州校区北区学生宿舍
日照科技文化中心
廊坊临空服务中心
文安文化中心
西安长陵博物馆
宜城四馆两中心

2016年赴斯里兰卡考察杰弗里·巴瓦建筑作品并参加研讨会
2016年参加首尔"中日韩书·筑"展
2016年应邀在韩国济州国际建筑论坛上演讲
2017年录制CCTV-1《开讲啦》节目
2017年赴韩国参加"首尔宗庙片区旧城改造"国际竞赛评审会
2017年赴美国加州大学伯克利分校作为访问学者为期半年
2017年被中国建筑学会任命为第四届亚太经合组织（APEC）建筑师中国监督委员会主任委员

2017年中国院方案组成立20周年"承"主题展开幕
2018年赴日内瓦会见丁肇中先生，并汇报了日照市科技馆方案
2018年参加第16届威尼斯国际建筑双年展香港百塔主题展
2018年应邀赴伯克利大学参加何镜堂院士展览开幕并发表演讲
2018年在郑州出席城市设计国际论坛并主持APEC建筑师中央理事会
2018年率中国建筑学会代表团赴韩国平昌参加中日韩三国交流会并发表演讲

2019年参加"未知城市：中国当代建筑装置影像展"
2019年接受人民日报社《环球人物》周刊专访
2019年出席香港建筑师学会两岸四地建筑设计论坛并做嘉宾演讲
2019年当选全国注册建筑师管理委员会主任委员
2019年参加首尔建筑与城市双年展
2019年主持第九届梁思成奖颁奖典礼
2019年出席巴黎国际日光研讨大会并代表中国代表团做主题演讲

2016 - 2017 | 2017 - 2018 | 2018 - 2019

中国大百科全书出版社改造 ·北京

万州三峡移民纪念馆 ·重庆

海口市民游客中心 ·海口
全国绿色建筑创新奖二等奖
全国优秀工程勘察设计行业一等奖
中国建筑学会建筑设计奖公共建筑一等奖、绿色生态技术一等奖

青海省图书馆、美术馆、文化馆 ·西宁
北京市建筑工程设计综合二等奖

北京西郊汽配城改造 ·北京

海口美舍河湿地公园生态科普馆 ·海口

北京外国语大学综合楼 ·北京

厦门中心 ·厦门

雄安市民服务中心企业办公区 ·雄安
全国绿色建筑创新奖一等奖
全国优秀工程勘察设计行业一等奖
中国建筑学会建筑设计奖公共建筑一等奖
中国建筑学会建国70周年创作大奖

首都师范大学行政及教学楼 ·北京
北京市建筑工程设计综合三等奖

敦煌市公共文化综合服务中心 ·敦煌

雄安设计中心 ·雄安
北京市建筑工程设计综合一等奖
中国建筑学会建筑设计奖公共建筑一等奖、绿色生态技术一等奖

天地邻枫创新产业园 ·北京

曲阜鲁能JW万豪酒店 ·曲阜
全国优秀工程勘察设计行业二等奖

2018中国（南宁）国际园林博览会 ·南宁
中国建筑学会建筑设计奖公共建筑一等奖、绿色生态技术二等奖

张家湾设计小镇智汇园 ·北京
北京市建筑工程设计综合三等奖

北京邮电大学沙河校区 ·北京

太原滨河体育中心 ·太原
中国建筑学会建筑设计奖公共建筑三等奖

昆山大戏院 ·昆山
北京市建筑工程设计综合一等奖

东北大学浑南校区图书馆 ·沈阳

深圳万科云城 ·深圳

招商银行深圳分行大厦
天津于家堡金融区03-15超高层办公楼
太原市图书馆
王府井步行街景观休息座区

昆山巴城影剧院改造
中国驻法国大使馆改造
北京雅昌艺术园区
敦煌世界地质公园雅丹景区游客中心
北京城市副中心行政办公区A2项目

北京经开·壹广场
太原幼儿师范高等专科学校
济南舜通大厦

Judge of International Competition, Valencia, Spain, 2007
Judge of International Competition, St. Paul, Brazil, 2007
Speech in Vienna, Austria, 2007
Judge of Taiwan Far East Architecture Awards, 2007
Liang Sicheng Award, 2007
International Architecture Exhibition, London, U.K., 2008
International Architecture Exhibition, New York, USA, 2008
Speech in Torino, Italy, 2008
International Architecture Exhibition, Paris, France, 2008
Co-Director of UIA Competitions Commission, 2008

Native Design published, 2009
Speech in Macau, China, 2009
UIA Board Meeting & Competitions Commission Meeting, Brazil, 2009
Solo Architecture Exhibition, Tianjin and Shenzhen, 2010
Architecture Exhibition, Yantai and Chengdu, 2010
Speech on 8th International Symposium on Architecture Interchanges, 2010
Judge of UIA 2011 Tokyo Competition, 2010

Be elected as Academician of China Engineering Academy, 2011
Board Meeting of UIA in Beirut, 2011
Speech in Chinese University of Hong Kong, 2011
Speech on UIA Congress Competitions Commission Meeting, Budapest, 2011
Judge of "Lin Tong / Heung Yuen Wai Joint Inspection Complex Building" International Competition, Hong Kong, 2011
"From Beijing to London – Contemporary Chinese Architecture Exhibition" and Speech in Newcastle University, 2012

2006 - 2008 | 2008 - 2010 | 2010 - 2012

Liangshan Nationality Culture Art Center, Xichang
■ ■ ■ ■

Han Meilin Art Gallery, Beijing
■

Visitors Center of Peach Blossom Valley, Mount Taishan, Taian

Library of Shandong University of Technology, Zibo
■ ■

Hongshan Site Museum, Wuxi
■

Zhejiang University Zijingang Campus Agriculture & Ecology Group, Hangzhou
■ ■

Business Office Area of Songshan Lake New Town, Dongguan

Plot B of the Inside-out, Beijing
■ ■ ■ ■

Ordos Dongsheng Sports Stadium, Ordos
■

BOBO International Office Building, Ningbo

Shandong Broadcasting & Television Center, Jinan
■ ■ ■

Beichuan Cultural Center, Beichuan
■ ■ ■

Wunushan Museum, Huanren
■ ■

The Legation Quarter, Beijing

The Embassy of China in South Africa, Pretoria
■ ■ ■

Multi-purpose Broadcasting Tower in Olympic Green, Beijing
■

Beijing Digital Press Information Center, Beijing
■

The Consulate General of China in Cape Town, Cape Town
■ ■

NO.3 Subsided Court of Olympic Green, Beijing
■ ■ ■

Reconstruction of CSEC Office Building, Beijing
■

Deyang Olympic Sports School, Deyang
■ ■

Complex Gymnasium of Beijing Foreign Studies University
Qinan Dadiwan Site Museum
Anhui Press Center
Tianjin The Ocean Paradise
Tai'an Ramada Plaza Hotel

Chengde City Planning Exhibition Hall
Jinan Quancheng Park People Fitness Center
Shanhaiguan Great Wall Museum Extension
Ren'ai College Library of Tianjin University
BPU Software Park II – Plot B
Office Building of Beijing Foreign Studies University
Library of Renmin University of China

The Chinese Museum of Women and Children
Dagukou Fort Ruins Museum
Fujian Broadcasting & Television Center

■ Architectural Society of China Creative Architecture Award ■ ARCASIA Award for Architecture Excellence Golden Medal ■ National Green Building Renovation Aw

Ten Years Cultivation: The Architectural Creation Exhibition of Cui Kai Studio 10th Anniversary, Beijing, 2013
Judge of the preliminary match of Far Eastern Architectural Design Award, 2013
Hired as Practical Professor of School of Architecture, Tsinghua University, 2013
Judge of the final match of Far Eastern Architectural Design Award, 2014
UIA Congress, Durban, South Africa, 2014
Speech on 16th Academic Exchange Meeting on Architecture from both Sides of the Strait, China, Taiwan, 2014

CUIKAI Studio renamed as the Land-based Rationalism D·R·C, 2014
Be director of the training program "Inside-out School", 2014
Speech on 14th World Triennial of Architecture, Sofia, Bulgaria, 2015
Be elected as academician of International Architecture Academy, Bulgaria, 2015
Leading a delegation to visit Society of Czech Architects, Prague, Czech, 2015

Speech in Hong Kong University, 2015
Judge of UIA Jiangxi Anji International Design Competition, 2015
Hired as Vice Chairman of Urban Design Experts Committee of the Ministry of Housing and Urban-Rural Development, 2016
IAA Council meeting, Sochi, Russia, 2016
Honorary member of Hong Kong Institute of Architects and Judge of the Graduate Design of the University of Hong Kong, 2016
Land-based Rationalism II published, 2016
"Towards A Critical Pragmatism: Contemporary Architecture in China" in Harvard Graduate School of Design, 2016

2012 - 2014 | 2014 - 2015 | 2015 - 2016

Chinese Hangzhou Cuisine Museum, Hangzhou

Khamba Arts Center, Yushu

Main Building of Tianjin University New Campus, Tianjin

Suzhou Railway Station, Suzhou

Mogao Grottoes Digital Exhibition Center, Dunhuang

Laboratory Building of Tianjin University New Campus, Tianjin

Nanjing University of the Arts, Nanjing

Library of Jiangsu Jianzhu Institute, Xuzhou

Site Museum of Ancient Tusi Castle, Yongshun

Kunshan Cultural Arts Center, Kunshan

Xi'an Dahua Model, Xi'an

Senior Department of Beijing No.35 Middle School, Beijing

CITIC Pacific Zhujiajiao Jin Jiang Hotel, Shanghai

No.4 Teaching Complex of Beijing University of Technology, Beijing

Lanzhou Planning Exhibition Hall, Lanzhou

Plot A, F of the Inside-Out, Beijing

Datong Museum, Datong

Kunshan Xibang Village Kun Opera School, Kunshan

CITIC Jinling Hotel, Beijing

Beijing Olympic Tower, Beijing

Kunshan Jinxi Zhujiadian Brickyard Reconstruction, Kunshan

BPU Software Park II – Plot E
Ordos Dongsheng Library
Penglai Ancient Ship Museum, Penglai
Western Returned Scholars Association, Beijing
BPU Software Park II – Plot C/D
Chongqing Guotai Arts Center
Library of Beijing Foreign Studies University

Ordos Sports Center
Linyi Grand Theatre
BDA One Center, Beijing
Inner Mongolia Broadcasting & Television Digital Media Center

Tourist Center of Ancient Tusi Castle, Yongshun
Library of Shahe Campus Central University of Finance and Economics
Changchun Planning Exhibition Hall & Museum

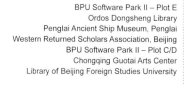
Beijing Best Project Design Award　　Architectural Society of China Grand Creation Award　　Others

2013年在北京举办"十年·耕耘"崔愷工作室建筑创作展
2013年赴上海参加台湾远东建筑奖评审会议
2013年被聘为清华大学建筑学院双聘教授
2014年出任台湾远东建筑奖决选评委
2014年赴南非参加世界建筑师大会
2014年赴台湾参加第三届海峡两岸建筑院校学术交流工作坊，暨第十六届海峡两岸建筑学术交流会并演讲

2014年崔愷工作室更名为"本土设计研究中心"
2014年主持开设"中间思库·暑期学坊"
2015年赴保加利亚参加第十四届世界建筑三年论坛
2015年荣获保加利亚国际建筑研究院院士称号
2015年带团访问捷克建筑学会

2015年赴香港大学建筑系讲学
2015年出任UIA江西安吉国际设计竞赛评委
2016年被聘为住房和城乡建设部城市设计专家委员会副主任委员
2016年赴俄罗斯索契参加IAA理事会会议
2016年被授予香港建筑师学会荣誉会员称号，并参加香港大学硕士生毕业评图
2016年《本土设计Ⅱ》出版发行
2016年参加哈佛大学设计学院"走向批判的实用主义：当代中国建筑"展览

2012 - 2014　　　　2014 - 2015　　　　2015 - 2016

中国杭帮菜博物馆·杭州
全国优秀工程勘察设计行业一等奖
中国建筑学会建筑创作银奖

康巴艺术中心·玉树
全国优秀工程勘察设计行业一等奖
中国建筑学会建筑创作金奖
中国建筑学会建国70周年创作大奖

天津大学新校区主楼·天津
全国优秀工程勘察设计行业二等奖
中国建筑学会建筑设计奖公共建筑类金奖

苏州火车站·苏州
中国建筑学会建筑创作金奖

敦煌莫高窟数字展示中心·敦煌
全国优秀工程勘察设计行业一等奖
中国建筑学会建筑创作银奖
中国建筑学会建国70周年创作大奖

天津大学新校区综合实验楼·天津
全国优秀工程勘察设计行业三等奖

南京艺术学院·南京

江苏建筑职业技术学院图书馆·徐州
全国优秀工程勘察设计行业一等奖
中国建筑学会建筑创作银奖
中国建筑学会建国70周年创作大奖

湖南永顺老司城遗址博物馆·永顺
全国优秀工程勘察设计行业二等奖
中国建筑学会建筑创作优秀奖、景观设计一等奖

昆山市文化艺术中心·昆山
全国优秀工程勘察设计行业二等奖

西安大华1935·西安
全国优秀工程勘察设计行业一等奖
中国建筑学会建筑设计奖建筑保护与再利用类银奖

北京三十五中学高中部·北京
北京市建筑工程设计综合二等奖

中信泰富朱家角锦江酒店·上海
全国优秀工程勘察设计行业二等奖

北京工业大学第四教学楼组团·北京
北京市建筑工程设计综合一等奖

兰州市城市规划展览馆·兰州
全国优秀工程勘察设计行业二等奖
中国建筑学会建筑设计奖二等奖

中间建筑A、F区·北京
全国优秀工程勘察设计行业一等奖
中国建筑学会建筑创作银奖

大同市博物馆·大同
中国建筑学会建筑创作银奖

昆山西浜村昆曲学社·昆山
中国建筑学会建筑设计奖公共建筑类银奖

中信金陵酒店·北京
全国优秀工程勘察设计行业一等奖
中国建筑学会建筑创作银奖

北京奥运塔·北京
全国优秀工程勘察设计行业二等奖
中国建筑学会建筑创作银奖

昆山锦溪祝家甸砖厂改造·昆山
全国优秀工程勘察设计行业一等奖
中国建筑学会建筑设计奖建筑保护与再利用类金奖
中国建筑学会建国70周年创作大奖

北工大软件园二期E地块
鄂尔多斯东胜图书馆
蓬莱古船博物馆
欧美同学会改扩建
北工大软件园C、D区
重庆国泰艺术中心
北京外国语大学图书馆

鄂尔多斯市体育中心
临沂大剧院
北京经开·壹中心
内蒙古广播影视数字传媒中心

湖南永顺老司城游客中心
中央财经大学沙河新校区图书馆
长春市规划展览馆、博物馆

2007年出任西班牙瓦伦西亚国际建筑竞赛评委
2007年出任巴西圣保罗国际竞赛评委
2007年在奥地利维也纳演讲
2007年出任台湾远东建筑奖评委
2007年获梁思成建筑奖
2008年于英国伦敦参加国际建筑展
2008年于美国纽约参加国际建筑展
2008年在意大利都灵演讲
2008年于法国巴黎参加国际建筑展
2008年担任国际建协国际竞赛委员会联席主任

2009年《本土设计》出版发行
2009年在澳门演讲
2009年赴巴西参加国际建协理事会和竞赛委员会工作会议
2010年分别于天津和深圳举办个人展
2010年分别于烟台和成都参加建筑展
2010年率团赴日本参加第八届亚洲建筑国际交流会并做主旨演讲
2010年出任国际建协2011东京世界建筑师大会设计作品评审会评委

2011年当选中国工程院院士
2011年赴贝鲁特参加国际建协理事会
2011年赴香港中文大学建筑系演讲
2011年参加在布达佩斯召开国际建协竞赛委员会工作会议并演讲
2011年出任香港"莲塘/香园围口岸联检大楼"国际竞赛评委
2012年赴伦敦参加"从北京到伦敦——当代中国建筑作品展"并在纽卡斯尔大学做演讲

2006 - 2008　　　　2008 - 2010　　　　2010 - 2012

凉山民族文化艺术中心 ·西昌
全国优秀工程设计银奖
全国优秀工程勘察设计行业一等奖
中国建筑学会建筑创作佳作奖
中国建筑学会建国60周年创作大奖

韩美林艺术馆 ·北京
全国优秀工程勘察设计行业二等奖

泰山桃花峪游客中心 ·泰安

山东理工大学图书馆 ·淄博
全国优秀工程勘察设计行业三等奖
中国建筑学会建筑创作金奖

无锡鸿山遗址博物馆 ·无锡
中国建筑学会建筑创作优秀奖

浙江大学紫金港校区农生组团 ·杭州
全国优秀工程勘察设计行业三等奖
中国建筑学会建筑创作银奖

东莞松山湖商务办公小区 ·东莞

中间建筑B区 ·北京
亚洲建筑师协会荣誉提名奖
全国优秀工程设计银奖
全国优秀工程勘察设计行业二等奖
中国建筑学会建筑创作佳作奖

鄂尔多斯市东胜体育中心 ·鄂尔多斯
全国优秀工程勘察设计行业二等奖

宁波BOBO国际办公楼 ·宁波

山东广播电视中心综合业务楼 ·济南
全国优秀工程设计银奖
全国优秀工程勘察设计行业一等奖
中国建筑学会建筑创作佳作奖

北川羌族自治县文化中心 ·北川
全国优秀工程勘察设计行业一等奖
中国建筑学会建筑创作优秀奖
中国建筑学会建国70周年创作大奖

五女山博物馆 ·桓仁
全国优秀工程勘察设计行业二等奖
中国建筑学会建国60周年创作大奖

前门23号 ·北京

中国驻南非大使馆 ·比勒陀利亚
全国优秀工程勘察设计行业一等奖
中国建筑学会建筑创作银奖
中国建筑学会建国70周年创作大奖

奥林匹克公园多功能演播塔 ·北京
全国优秀工程设计铜奖

北京数字出版信息中心 ·北京
全国优秀工程勘察设计行业二等奖

中国驻南非开普敦领事馆 ·开普敦
全国优秀工程勘察设计行业一等奖
中国建筑学会建筑创作银奖

奥林匹克公园下沉花园3号院 ·北京
亚洲建筑师协会优秀建筑设计金奖
全国优秀工程设计银奖
全国优秀工程勘察设计行业一等奖

中国神华办公楼改扩建 ·北京
北京市建筑工程设计综合三等奖

德阳市奥林匹克后备人才学校 ·德阳
全国优秀工程勘察设计行业二等奖
中国建筑学会建筑创作金奖

北京外国语大学综合体育馆
秦安大地湾遗址博物馆
安徽出版集团大厦
天津海河新天地
泰安东尊华美达酒店

承德城市规划展览馆
济南泉城公园全民健身中心
山海关长城博物馆扩建工程
天津大学仁爱学院图书馆
北工大软件园二期B地块
北京外国语大学教学办公楼
中国人民大学图书馆

中国妇女儿童博物馆
大沽口炮台遗址博物馆
福建广播电视中心

2016 - 2017	2017 - 2018	2018 - 2019
Investigation & Seminar of Geoffrey Bawa Design Works, Sri Lanka, 2016 China-Japan-ROK Architectural Books Exhibition, Seoul, South Korea, 2016 Speech on Jeju International Architecture Forum, Jeju, South Korea, 2016 CCTV-1 talk show *VOICE*, 2017 Judge of Jongmyo District Renovation International Competition, Seoul, South Korea, 2017 Visiting Scholar of the University of California at Berkeley for 6 months, 2017 Appointed as Director of the Forth Chinese Supervision Committee for APEC Architects, 2017	"Inheritance", Exhibition of the 20th anniversary of CADG Creative Group, 2017 Meeting with Mr. Samuel Ding and Report of Rizhao Science Museum, Geneva, Switzerland, 2018 Hong Kong Pavilion Theme Exhibition of the 16th Venice International Architecture Biennale, Venice, Italy, 2018 Speech on the opening ceremony of Academician He Jingtang Exhibition in Berkeley University, California, USA, 2018 Attending International Forum of Urban Design and be chairman of the Central Committee of APEC Architects, 2018 Speech on China-Japan-ROK Seminar as the head of ASC Delegation, Pyeongchang, South Korea, 2018	Unknown City: Installation and Imagery on Chinese Contemporary Architecture, Shenzhen, 2019 Interview of *Global People*, a magazine affiliated with *People's Daily*, 2019 Speech on the Hong Kong Institute of Architects Cross-Strait & Four Places Architectural Design Forum, 2019 Elected as the Chairman of National Board of Certified Architects, 2019 Seoul Architecture and Urban Biennale, 2019 Presided over the 9th Liang Sicheng Award Presentation Ceremony, 2019 Speech on International Daylight Conference as the head of Chinese Delegation, Paris, France, 2019

 Renovation of Encyclopedia of China Publishing House, Beijing

 Wanzhou Three Gorges Migrant Memorial Hall, Chongqing

 Haikou Citizen & Tourist Center, Haikou
■ ■ ■

 Qinghai Library, Art Gallery and Cultural Hall, Xining
■

 Renovation of Beijing Western Suburbs Auto Parts Mall, Beijing

 Ecological Science Museum in Haikou Meishe River Wetland Park, Haikou

 Comprehensive Teaching Building of Beijing Foreign Studies University, Beijing

 Xiamen Center, Xiamen

 Xiong'an Civic Service Center Enterprise Office Area, Xiong'an
■ ■ ■ ■

 Administration & Teaching Building of CNU, Beijing
■

 Dunhuang Public Culture Comprehensive Service Center, Dunhuang

 Xiong'an Design Center, Xiong'an
■ ■

 Tiandi Linteng Green Inaustrial Park, Beijing

 JW Marriott Hotel, Qufu
■

 2018 China(Nanning) International Garden Expo, Nanning
■

 Zhihui Garden of Zhangjiawan Design Town, Beijing
■

 Shahe Campus of Beijing University of Posts and Telecommunications, Beijing

 Renovation of Taiyuan Binhe Sports Center, Taiyuan
■

 Kunshan Grand Theater, Kunshan
■

 Hunnan Campus Library of Northeastern University, Shenyang

 Vanke Cloud City, Shenzhen

Shenzhen Branch of China Merchants Banks 03-15 Super High-Rise Office Building, Yujiapu Financial District, Tianjin Taiyuan Library Wangfujing Pedestrian Street Landscaped Seating Facility	Renovation of Kunshan Bacheng Cinema Renovation of the Embassy of China in France Beijing Artron Art Park Dunhuang World Geopark Yadan Landform Area Tourist Center Project A2 in the Administrative Area of Bating Sub-Center	BDA One Plaza, Beijing Taiyuan Preschool Teachers College Jinan Shuntong Building

Award Presenter of the Architectural Society of China Grand Creation Award for the 70th Anniversary of the Founding of the People's Republic of China, and won 8 awards, 2019
"The Eye of the City" Theme Exhibition of Shenzhen and Hong Kong Urban Architecture Biennial, 2019
What does the Pandemic Reminds Us? in *Architectural Journal*, 2020
Judge of the Architecture and Structure Design Competition for College Students of Belt and Road Architectural University International Consortium, 2020
Final Judge of Architecture China Award, 2020

"Evolution of Cities" Exhibition & Forum, 2020
Opening of "D·R·C Seasonal Exhibition", which presents three scheme exhibitions per year exhibiting research achievement of D·R·C, 2021
Interview of "The Architects" program, Phoenix Chinese Channel, 2021
Green Architecture Design Guidelines published, 2021
China Pavilion Theme Exhibition of the 17th Venice International Architecture Biennale, Venice, Italy, 2021
Postgraduate program "Introduction to Practicing Land-based Rationalism" in Beijing University of Civil Engineering and Architecture, 2021

First Prize of National Prize for Progress in Science and Technology, 2021
Research on Regional Climate Adaptable Green Public Building Design Series, 2021
As a panelist in the China-Brazil Architecture Forum, Beijing, 2021
Opening of "CADG History Exhibition Gallery" for the 70th Anniversary Celebration of CADG, 2022
Land-based Rationalism – Urban Renovation Itinerant Exhibition, Chongqing, 2022

2019 - 2020   2020 - 2021   2021 - 2022

 China Pavilion at The International Horticultural Expo 2019, Beijing

 Suining Song Porcelain Culture Center, Suining

 Main Exhibition Hall of Tianfu Agricultural Expo Park, Chengdu

 Beijing-Tibet Communication Center, Lhasa

 Rongcheng Youth Activity Center, Rongcheng

 Qingdao SCODA Pearl International EXPO Center, Qingdao

 The Railway Brigades Memorial, Zaozhuang

 Zhangjiagang Jingang Culture Center, Zhangjiagang

 Chongli Center, Chongli

 Baoji Culture and Art Center, Baoji

 Main Pavilion and Autograph Collection in Jiangsu Garden Expo, Nanjing

 Chongqing Planning Exhibition Hall, Chongqing

 Nanxun Planning Exhibition Hall, Nanxun

 Future Garden of Jiangsu Garden Expo, Nanjing

 Children-friendly Center of Xuzhou Garden Expo, Xuzhou

 Rizhao Science Museum, Rizhao

 School of Urban Design, Wuhan University, Wuhan

 Binhu High Middle School, Huzhou

 Xiong'an Railway Station, Xiong'an

 Small Land of Peach Blossoms, Kunshan

Sinovac Tianfu Factory Renovation, Beijing

Weihai Library & Group Art Hall
Baoding Guan Hanqing Grand Theater
Jingdezhen Jinrong Foreign Language School
Chengdu CRRC City Star, Phase I
Beijing Firm Academy New Camps in Huairou
Teaching & Service Group of Beijing Foreign Studies University West Campus

Yiyang Civic Culture Center & Government Affairs Center
North Building of Architecture School, Tianjin University
Daxing Airport north line expressway Langfang Tall Gate
Stadium of Yan'an National Fitness Sports Center
Yulin Dongsha Culture and Sports Hall
Zhoushan Culture Center
Kunshan Western Medical Center, Phase I
Xiangyu Group Office Building

Jingdezhen Vocational University of Art, Phase I
Tsinghua Shenzhen International Graduate School, Phase I
Student Dormitory in North of Tongzhou Campus, Renmin University of China
Rizhao Science and Culture Center
Langfang Airport Service Center
Wen'an Culture Center
Xi'an Changling Museum
Four Halls and Two Centers in Yicheng

1957年出生于北京

景山学校就读小学、初中、高中
北京平谷县华山公社麻子峪插队

1984年毕业于天津大学建筑系获硕士学位,师从彭一刚教授
1984年进入建设部建筑设计院
1985-1989年任职华森建筑与工程设计顾问有限公司
1989-1997年任职建设部建筑设计院
1997年任建设部建筑设计院副院长
2000年任中国建筑设计研究院副院长、总建筑师

1999-2008年任国际建协副理事
2001年在马来西亚吉隆坡演讲
2004年在韩国釜山演讲
2004年出任韩国釜山国际竞赛评委
2004年于中国台北参加国际建筑展
2000年任中国建筑学会副理事长
2000年当选全国工程勘察设计大师
2003年获法国文学与艺术骑士勋章
2003年崔愷工作室成立

2005年在韩国首尔演讲
2005年在新加坡演讲
2005年于深圳参加双年展
2006年在斯里兰卡科伦坡演讲
2006年出任香港建筑学会年奖评委
2006年在荷兰代尔夫特演讲

1984 - 2000　　　　　　2000 - 2004　　　　　　2004 - 2006

阿房宫凯悦酒店 ·西安
建设部优秀建筑设计二等奖

水关长城三号别墅 ·北京

殷墟博物馆 ·安阳
亚洲建筑师协会优秀建筑设计金奖
中国建筑学会建筑创作优秀奖
中国建筑学会建国60周年创作大奖

蛇口明华船员基地 ·深圳
建设部优秀建筑设计三等奖

清华科技创新中心 ·北京
全国优秀工程设计铜奖
建设部优秀建筑设计二等奖

德胜尚城 ·北京
全国优秀工程设计铜奖
全国优秀工程勘察设计行业二等奖
中国建筑学会建筑创作佳作奖

丰泽园饭店 ·北京
全国优秀工程设计铜奖
建设部优秀建筑设计二等奖
中国建筑学会建国60周年创作大奖

中国城市规划设计研究院办公楼 ·北京
中国建筑学会建筑创作佳作奖

首都博物馆 ·北京
全国优秀工程设计银奖
全国优秀工程勘察设计行业二等奖
中国建筑学会建筑创作佳作奖
中国建筑学会建国60周年创作大奖

外交部怀柔培训中心 ·北京

富凯大厦 ·北京
全国优秀工程设计银奖
建设部优秀建筑设计二等奖
中国建筑学会建筑创作优秀奖

拉萨火车站 ·拉萨
全国优秀工程设计金奖
全国优秀工程勘察设计行业二等奖
中国建筑学会建筑创作优秀奖
中国建筑学会建国60周年创作大奖

外语教学与研究出版社办公楼 ·北京
全国优秀工程设计铜奖
建设部优秀建筑设计二等奖
中国建筑学会建筑创作优秀奖
中国建筑学会建国60周年创作大奖

民航总局办公楼改造 ·北京
建设部优秀建筑设计三等奖

大连软件园9号办公楼 ·大连
中国建筑学会建筑创作佳作奖

现代城高层公寓 ·北京
建设部优秀建筑设计三等奖

雅昌彩印大厦 ·北京
全国优秀工程设计银奖
建设部优秀建筑设计二等奖
中国建筑学会建筑创作佳作奖

大连软件园软件工程师公寓 ·大连
中国建筑学会建国60周年创作大奖

北京外国语大学逸夫教学楼 ·北京
全国优秀工程设计银奖
建设部优秀建筑设计二等奖

外研社大兴国际会议中心 ·北京
中国建筑学会建筑创作佳作奖

大连软件园10号、11号办公楼 ·大连
全国优秀工程勘察设计行业二等奖

　　　　　　　　　　　　　　　中国建筑设计研究院办公楼改扩建　　　　　　　　大连软件园8号办公楼
　　　　　　　　　　　　　　　　　　　　　宁波天一家园　　　　　　　　　　　　　西安紫薇山庄
　　　　　　　　　　　　　　　　　　　　　北京康堡花园　　　　　　　　　　天津开发区高校园配套公建区一期
　　　　威海中信金融大厦　　　　　　　　　成都东方明珠花园　　　　　　　　　　浙江大学生命科学学院
　　　外语教学与研究出版社印刷厂改造　　　　北京蓝堡国际公寓　　　　　　　　　　唐山国际会展中心
　　　　　　北京现代城幼儿园　　　　　　　天津开发区职业技术学院　　　　　　　　株洲体育中心
中国石油大学(华东)东营校区图书馆及饮食服务中心　　天津科技大学泰达校区

图书在版编目（CIP）数据

本土设计. Ⅲ = LAND-BASED RATIONALISM Ⅲ / 崔愷著. -- 北京：中国建筑工业出版社，2023.10

ISBN 978-7-112-29190-8

Ⅰ. ①本… Ⅱ. ①崔… Ⅲ. ①建筑设计—研究—中国 Ⅳ. ①TU2

中国国家版本馆CIP数据核字(2023)第180921号

策划统筹：张广源　傅晓铭
特邀编辑：刘爱华
英文翻译：谭雅宁　任　浩　黎晓晴
美术编辑：田歆颖

责任编辑：欧阳东　陈　桦　杨　琪
责任校对：党　蕾

本土设计Ⅲ
LAND-BASED RATIONALISM Ⅲ

崔愷　著
*
中国建筑工业出版社出版、发行（北京海淀三里河路9号）
各地新华书店、建筑书店经销
北京雅昌艺术印刷有限公司制版
北京雅昌艺术印刷有限公司印刷
*
开本：880毫米×1230毫米　1/16　印张：22　插页：29　字数：847千字
2023年10月第一版　　2023年10月第一次印刷
定价：298.00元
ISBN 978-7-112-29190-8
　　（41893）
版权所有　翻印必究
如有内容及印装质量问题，请联系本社读者服务中心退换
电话：(010) 58337283　QQ：2885381756
(地址：北京海淀三里河路9号中国建筑工业出版社604室　邮政编码：100037)